The Healing Power of Chlorophyll
from Plant Life

by BERNARD JENSEN, D.C.

Edited by
Leslie Goldman

MAGIC SURVIVAL KIT — Book One

**Published by
Bernard Jensen Enterprises
24360 Old Wagon Road
Escondido, California 92027**

DEFINITIONS:

GREENS/CHLOROPHYLL: GREEN is one of the greatest rejuvenating colors, found in the greatest of the rejuvenating plants. The GREEN in plant life repairs human tissue when edible GREEN vegetables are taken internally or used externally. The CHLOROPHYLL green pigment is the life-force of the plant, and this force provides life-giving materials for the human body.

SEEDS: The next generation is in the seeds, and herein lies the greatest storehouse of nutrients Nature offers. Nuts, grains, and legumes are all SEEDS. Nature refines herself as she goes on; the next generation should always be better than the last.

BERRIES: The most natural of all the foods we have today, BERRIES to a great extent have escaped the "watering-down process" of hybridization, the fate of most of our vegetables and fruits. Berries are loaded with seeds as they should be and each seed radiates life to every bit of juice around it. The removal of seeds from a fruit upsets the glandular balance in that fruit and in the man who eats it.

SPROUTS: As perfect a structural being as it will ever be, SPROUTS represent the beautiful child born to the mother. The first five days of the sprouted seed represent the first nine months in the life of the child. A vast array of seeds can be sprouted as vital survival food for man.

SURVIVAL: The word "SURVIVAL" should not cause fear, it means FREEDOM. SURVIVAL is the ability to take care of yourself successfully with the least amount of "dis-ease" in this day and the next through lean times and times of plenty. All that keeps pestilence and famine from view today is a thin bandage — an excellent cosmetic treatment on our crops which masks the truth. Today is a SURVIVAL day demanding as much SURVIVAL planning as any future day. Each bite of food we eat contributes to our personal health SURVIVAL or lack of it.

MAGIC: At one time, priests in Mexico were called magicians. They were not magicians, in the modern sense of the word; they just happened to know the survival laws — how to read Nature's manual of rules. These 'magicians' watched holes in the pyramids. When the shadows from the sun reached a certain point, they advised the people to plant their crops immediately. If the people ignored the advice of the priests, their crops were poor and the harvest meager. Actually, there was no MAGIC to it at all, only an understanding of Nature.

SURVIVAL LAWS: SURVIVAL LAWS are the time-tested rules which exist in Nature. Despite all the preparations man makes for his future, SURVIVAL LAWS are all that stand between him and his destiny. Our economic structure is based on a concept of insurance. Learning the SURVIVAL LAWS and practicing them is the best insurance policy you can carry.

ABOUT THIS MAGIC SURVIVAL KIT. . .

As an owner of health ranches, I have lived close to Nature for more than 45 years, and I have learned to think in a survival way. We have had to struggle through loss of gas, water, and electricity. Now loss of these resources is no longer a problem to me. I can remember traveling to town on horseback to pick up supplies when floods washed out our roads.

I can take care of myself, but I am thrown into a panic watching our guests from the city attempt to cope with these "world-shaking" losses. The average person would be lost without money in his pocket and insurance for a "rainy day". Yet there may come a day when shortages will occur not only in money but pants and pockets! Banks may be unable to supply resources to keep man well-fed. A $1,000 bill may not buy an ear of corn.

We must take care of ourselves, yet we have gotten far from Nature. The average person in the city has never watched a carrot planted, harvested, or stored away in a root cellar. We have become apartment house dwellers, no longer children of the sun, but of the drugstore, supermarket; of pavements and streets we call "the jungle".

Nature balances life by having days of plenty and days of shortage. We have had "ups and downs" in the past, and we will have them again in the future. Changes are bound to come in man's system while Nature continues on.

Man Can Plan Ahead

What will you do in a time of crisis? What will you do if a disease comes upon you? Because of his mind, man can look ahead and help plan his destiny; the lights are going out everyday through lack of knowledge!

The wise man knows how to buy foods, how to grow foods, how to dehydrate foods, and then how to revive them for future use. He can prepare foods for storage with a minimal loss of vitamins and minerals.

The wise man knows that he is his brother's keeper! It will not be a matter of giving him money only; his real need will be food in time of crisis.

We should meet our survival moment — each and every moment — in the best of health. A working knowledge of the basic foods in this *Magic Survival Kit* will help meet this moment in good health. By using knowledge of the survival foods from day to day and in a time of crisis, we can successfully endure any lean periods which may occur.

A Sharing of Ideas

After more than 45 years working as a nutritional advisor, health counsellor,, chiropractor, and lecturer — all coming under the heading of "Doctor," the day has come to pioneer new fields; to grow my own garden. Now I must share ideas from my own survival studies.

My survival ideas have taken shape over a period of years. I have come to think in terms of survival, not out of fear but out of a desire to make life more fulfilling. Dying is easy. It's living that's hard.

In Book One of this *Magic Survival Kit* I offer some basic notes describing Nature's ways, some insights into the condition of man and his age, and an introduction to a crucial survival food, Chlorophyll, Nature's greatest healer.

Book Two will present other basic survival foods — the berries, seeds (including nuts and legumes), sprouts, and foods from the "land of milk and honey". These foods are crucial to your health survival.

In Book Three, I will share my personal survival plans with you — ideas I have formed from my work in Nature's healing realm. This book will include methods from bass farming to the rapid growing of survival foods through use of "pinkhouses." These are ideas I am now implementing in my own survival gardens.

If you are seriously interested in survival, then I want to bring this kit to you. Your life is your own, but I believe your life is your own to be *lived* as it can be lived. If some students are sent toward the days ahead with an inkling of knowledge because of what I share here, then my purpose will have been served.

Bernard Jensen
Hidden Valley
March, 1977

So much of Nature as he is ignorant of, so much of his own mind does he not yet possess. And in fine, the ancient precept, "Know Thyself," and the modern precept, "Study Nature," become as last one Maxim."

The American Scholar
Ralph Waldo Emerson

TABLE OF CONTENTS

Part I

Nature: Key to Survival

Some Insights about Man, his Age,
and the Natural Law that Governs Both

MAGIC SURVIVAL KIT — Book One

CHAPTER ONE

THE MANUAL THAT GOT AWAY

Your new Chevrolet is waiting for you down at the car dealer. When you open the door, you will find a manual of operation on the front seat. It tells you how to start the car, how to steer, when to oil, and how to grease it. When you buy a food blender, with it you get a manual that explains how the machine runs. It cautions you against putting in too much food. If you follow the instructions both pieces of machinery will "Survive" a long time.

When man came into this world, a "manual" was placed along side of him, but somehow or other, it got lost. Many of us have been trying to read this manual nevertheless. We try to find out how to grease and oil the parts and how to keep the joints working and the hinges in operation.

Some men are busy installing new hinges and parts. Man is getting this transplanting business down to a really fine art these days, but somehow or other, trying to put new organs in an old body doesn't work.

We send old cars out to the junk yard, and they stay there. They don't seem to come back new. But here is the loveliest thought I can share with you: YOU CAN GO TO THE JUNK YARD WITH THIS BODY AND COME BACK NEW!

3

The Principle of Nature Cure

But to do so, you must follow this manual — the Laws of Nature — to the letter. The manual - your birthright - was written by a divinely inspired architect and chemist. The rules have been written in Nature.

Nature is organized so man can live forever. However, the weak are not meant to live on. Nature takes away the sick. The man who is weak must become strong. This is done by following the law. If you can't keep your body clean, Nature will be taking you out before long. Her only interest is survival of the fit.

Our Body is Self-Rejuvenating

Talk about health building - patients who have found Nature's path often pass through my Health Ranch again. Years later, I can see the changes they have made. Sometimes it takes a year or four or five years or even longer depending on how much they were broken down, but with knowledge we can take our little machine back from the scrapheap. Our body is self-repairing, self-rejuvenating, and self-building.

Survival: Taking Care of Yourself

When we talk about survival it is not a matter of the world out there coming to an end that should concern us. The world we must take care of is our own body; we are a piece of mud as much as the earth is. I see many people's days coming to an end because they haven't learned how to take care of their own 10-acre piece of ground.

Survival isn't a matter of waiting for a crash, waiting until there are shortages of money, food, fuel, and energy. The shortages that concern us are the CHEMICAL ELEMENTS AND ENERGY THAT FLOW THROUGH US.

Right this moment we are in an age when we must consider our personal health survival. THIS IS OUR SURVIVAL MOMENT.

All my life I have been living in a survival world. I have tried to bring lives to good health, tried to help my patients overcome disease, tried to help overcome bad habits and therefore prevent disease.

On a physical level, I repeat, survival is in our chemical elements, and mentally, in having a good personal philosophy. Our attitude can destroy the finest physical regimen. We can create more damaging body acids by misthinking than we can neutralize with all the alkaline-forming broths we can take.

We must keep from chaos, go through life in a harmonious way, and lead a life where we are elevated to the highest thoughts possible. No one should be condemned or banned, but rather well-nourished, well-thought of, blessed, lifted. We are here to do something good for one another.

Only Nature Cures

I am a Nature boy, and I believe in Nature. I believe that without her, we're not going anywhere. I believe Nature cures and only she cures, but Nature needs an opportunity. Nature is very slow, but she is just and thorough. I believe we are natural people. We are children of Nature and of a divine spirit - both one in the same.

The closer we come to Nature and follow her laws, living within cold, heat, sun, air, water, the closer we will come to health.

CHAPTER TWO

LIFE DEPENDS ON THE CHEMICALS IN US

If only we would look to the outside of our body — look to all the garden! Our bloodstream is lying out there right now in fruits and vegetables. It's out there in nuts, seeds, sprouts, and berries!

The body is made of chemical elements, the same as found in the "Dust of the Earth." We lean on plantlife to transfer these chemical elements from the soil to our body. Plant life transforms and transmutes these chemical elements from a lower to a higher biochemical existence for man's use.

We depend on this "plant" go between. It is difficult for our body to take minerals directly from the earth and use them to benefit our body nutritionally.

We try to get our health back. We never can unless we know the chemical elements and the chemical story. If we could read the symptoms, and analyze them in terms of chemical shortages, then we could replace these chemical elements.

SURVIVAL LAW NO. 1

Every Disease is a sign of a lack of certain chemical elements. No disease can exist unless there is a lack of chemical elements in the body.

Coming Up Short

Today we may have a body which is chemically in balance; however, the rigors of these modern times place extra strain on us.

1. You have an accident and break five bones. You possibly could have had enough calcium to take care of everyday needs, but when the accident occurs your body demands extra calcium.

2. You may give birth to a child. Before the child was born, you may have had good teeth. In order to have a child who is as perfect as possible, Nature is going to see that the child has a good start in life. In doing so Nature may rob the mother who may lose calcium from the teeth, or her hair may fall out. Many other conditions may occur.

3. There are jobs that act as robbers. They demand a certain part of your body to get that job going. Foods that can keep you healthy while working at that job are needed.

4. Your marriage also can wear out certain chemical elements in your body, and these conditions can be serious. Certain chemical elements can be worn out and you become fat. Chemical elements can be worn out and cause you to become so thin you will look like a stringbean!

YOU CAN BURN OUT YOURSELF PHYSICALLY, MENTALLY, OR SPIRITUALLY. MANY PATIENTS WHO COME TO ME TO GET WELL ARE BURNED OUT.

We are constantly in motion with the "Dust of the Earth," and we constantly must replace the chemical elements we use. As these chemical elements flow through all Nature, eventually they must flow through us. We belong to the garden. These foods were created and built for us, and we were built to have them. We ought to get together!

The secret of food chemistry is to replace poorly functioning body chemistry and to put back these shortages. The doctor who works

---

SURVIVAL LAW NO. 2

Our body molds to the foods we eat and to the foods we do not eat.

with Nature can interpret symptoms of shortages and suggest foods which will make up for these chemical shortages.

Food can really do things. I can't believe food can make changes in the body the way it does. When I was in Germany I saw a horse fifteen hands high — Quite a tall horse.

By taking away foods from this horse in nine generations the horse's family would be a foot-and-a-half high, yet through proper feeding, members of the same family could be built to good size horses again.

When we look at the forms of healing and "doctoring" today, I haven't a single experiment to show that if there is one way better than another, it is the way of Nature.

Building a New Body

I'm interested in correction. If you want a new body you can work for it. New skin grows on the palm of your hand every 24 hours. If you want new tissue in place of the old, you have to live a better life. The health principle in your body cannot be elevated unless you live on cleaner and more wholesome foods.

The life principle that flows through us will do the best it can on coffee and doughnuts. When we try to get the good out of poor materials, we only make poor bodies. This life principle will do a much better job if we eat salads and some of the fresh clean foods we were meant to eat. If you are satisfied with the same old kind of tissue, then keep living on the foods you are eating now.

How Elements Come to Us

If Nature's manual seems distant, don't despair, for even in ancient times man missed the whole story. In the ancient scriptures it tells us "There is brass and iron for man's use," yet there has never been an element called brass. While men in those days may not have been the skillful "scientists" of today, they did know that "something" came from the soil. These vital elements - which they called brass and iron - were transferred from the soil to plant life when they came into our life as edible grasses, herbs, and the pulse of the field - vegetables.

CHAPTER THREE

THE NEED FOR INSTRUCTION

We can be poorly instructed. Nature's Manual hasn't been found by too many people. What has been coming out of man's mind because of poor instruction has been disastrous and destructive to himself and others.

The uninstructed person today is eating soda crackers. The other day I watched a mother giving her asthmatic child a candy sucker. Here was an asthmatic child receiving asthmatic causing food. This was something to think about.

Another lady bought 1000 aspirin because she was getting them for a bargain — $.49! "Why are you buying a 1000 aspirin?" I asked. She was buying 1000 aspirin for her 1000 anticipated headaches. She was not living, she was dying to get well. She was dying to be comfortable.

We are trying today to cure rather than prevent a disease and dis-harmony, including many a doctor.

The Need for Teachers

No one wants to go into your home and tell you something about the canned foods you are eating. No one wants to go into your home and tell you not to smoke. Our government tells us smoking is bad, but many haven't changed their habit.

I believe the average person today wants to do the right thing, but I don't believe he knows what to do. The average person is seeking and questioning and hungry for some good answers.

Without knowledge and vision we perish.

I see a short time ago, doctors found that hominy used by the Mexicans was causing ulcers of the stomach. This hominy is made through a lye process which contributes to the irritation. These doctors wouldn't do anything about this use of hominy because "hominy had been in the Mexican culture for years." These doctors did not want to disrupt that culture. I would.

Aren't we honest enough, square enough? Don't you know that if something is causing trouble, it is time to stop it. It is time to stop a lot of things and start a lot of new things.

Here we are, built in the image of perfection, yet look at all the thinking that has brought so many imperfect foods into existence today. We have made these things, and we have delivered them to human beings. We know many times they are causing harm to our fellow, yet we "cannot" stop it!

Divine Order

Plant life is here for a purpose, although not only for our purposes. You can't name a thing in this universe which doesn't fit perfectly in the order of things. This order goes right down to rats, mice, grasshoppers, parasites, and bugs. These are Nature's undertakers. Nature needs them to have good control. We haven't learned about the role of vegetation; we haven't learned what is good for us. We fail to realize how much we, too, are a part of the cycle of life. We seldom realize that anything that has ever lived has to go back to the soil again - this includes man himself.

Looking beyond Nature herself, we find her laws have passed to us written in many, many books, recorded there by many sages. They pass to us in the teachings of Moses, in the works of Confucius, Buddha and the Christ.

Yet, it takes no sage or anyone religiously affiliated to let us know that we live in a realm of cosmic action and reaction, and that order balances the confusion which some men see everywhere. There is a Creator who regulates and put things in good order.

Our Body Order

As we put things "in order" we find that human beings are orderly organized machinery, so well-constructed that a look within reveals "an instrument of a million strings" - 27,000 of them coming from the ears to the brain and another 500,000 from the eyes to the brain - and look at the strings in the other parts of the body!

Look at this body! It's a work of art. No improvements can be made on the basic design and function. It was divinely put together and it is beautiful and wonderful. It is for us to see this beauty in ourselves and in other people.

CHAPTER FOUR

THIS AGE WE LIVE IN

Man, who came out of the Garden of Eden, has really made a mess of things! Once he got started, he continued to make a mess, and he doesn't seem to be making things better.

When 92% of the people in this country are sick by government figures, here is a definite sign that most people do not know how to live right and that most doctors are not teaching people how to live well. It is a sign that we do not practice preventive medicine aimed at keeping us out of trouble. We have institutions today to take care of sick people rather than well people. We have hospitals to match our sickness.

We find cancer today accounts for 16% of our death rate in the United States. Over a period of years no matter how vigorously treated or diagnosed, four out of five cancer victims are going to die.

We find we are not having "too good" results curing this disease.

12

There has been a 1,400% increase in lung cancers in men in the last 40 years, and we find lung cancer reaching to epidemic size. Among women with breast cancer, just as many die today as did thirty years ago. One out of four people walking the streets of Seattle has a cancer and doesn't even know it, statistics tell us.

A Dr. Hanna, from the Oakridge laboratories, showed a short time ago that by injecting a TB anti-virus injection into guinea pigs, there was a great surge of white blood cells that would offset this TB injection. Practically all the guinea pigs were cured of skin cancers by doing so.

When information leaked out to the newspapers that a breakthrough for cancer had been found, the headlines were bigger than those printed when peace is declared during war time. And that was just for a cure for cancer in animals!

When news of this breakthrough was printed, thousands of animals were shipped to this Oakridge hospital! Someone even sent a horse all the way from Australia. We don't realize how rampant cancer is these days.

We would feel wonderful if we could see a cure for cancer for man, yet we go right on living in this carcinogenic age. More than 1,000 chemicals have been found to cause cancer in animals. We find these carcinogens used in additives to our foods. We also find them in our air, in pesticides, and injected into human beings.

Prevention is Needed

If we don't stop some of the factors causing cancer, what's the use of trying to implement a cure? Here we are, producing these diseases with the left hand and trying to stop them with the right. Why not get both hands going in the right direction?

We know what produces a lot of these cancers, and it's criminal to see what we are subjecting human beings to. Sometimes I don't know what to do myself. I feel like a little pebble on the beach, trying to say something.

Here DDT was taken off the market because it produced cancer in animals, and here it is back again in the animals and on our market shelves. There was such a storage of DDT that the manufacturers just couldn't see throwing it away! They can't throw it away because the bugs have been coming so fast that the farmers don't know how to handle them.

How Soil Chemistry Works

Little do we know that if we had a pure and clean soil and started replenishing the earth like the good book tells us to do, we wouldn't have these bugs eating up the crops. We must realize that NATURE'S Undertaker - the germ life, only eats crops that are not fit for our consumption in the first place!

As long as we have foods which aren't healthy, we always will have bugs and germlife on them.

Germlife is necessary. Germs have been created to bring the cycle of life into good order, to bring the unhealthy plant into the soil again. If the new vegetation has a good start the second time around, it will come up as vegetation for our use, and then after, return again to the soil.

The Wheel of Health

This is the cycle of life. This is the wheel of health, and we are part of this wheel of health.

We don't realize that 90% of us have tuberculosis germs in the throat, but that doesn't mean 90% of us have tuberculosis. But you can have a good time getting tuberculosis if that's your pleasure. All you must do is wear yourself down. Become an unhealthy "plant" and Nature will return you to the soil, and quickly!

CHAPTER FIVE

ABOUT OUR FOODS

Sometimes people ask me, "What depreciation in vegetables and fruits can we expect because these produce are grown with artificial fertilizers and sprays which the growers must use to keep Nature's undertakers off?"

I tell them first about organically grown fruits and vegetables which also can have a lack of mineral elements as well as that produce grown by artificial means. Organically grown produce may lack the minerals because these elements weren't in the compost - the material used for fertilizer. It's not just a question of something being "organic" - meaning grown by natural means. Most importantly, in order to be healthy, produce must grow on soil which is not only "organic" but pure, whole, and natural.

Just as we produce foods "short" in necessary elements, so we too become short when we eat these foods. We find also that not only can the foods we eat be short in chemical elements, but it is the accumulations of these dastardly man-made fertilizers which also can cause problems.

We create problem-foods when we grow them with unnatural fertilizers and sprays. These substances accumulate in the plant. We end up with a "drugged" food.

SURVIVAL LAW NO. 3

We come to this simple truth: We must REPLENISH the earth with everything that ever grew on this earth. Everything must come back as fertilizer for all the soil. When "All" comes back, we have a wholeness which will contain all the chemical elements needed for health. No shortages will occur in the man who partakes of food grown in this soil.

When we create a drugged food, we create a food that has an accumulative activity when it enters the human body. These sprays, pesticides, and inorganic fertilizers will eventually get to the tissues in our body and cause a disturbance. It's these substances that are found to be carcinogenic in most cases.

It's the addition of these substances that gets into our livestock. While the first piece of meat we eat isn't going to cause problems, the accumulative effect of many pieces of meat containing these additions will.

Our Carcinogenic Environment

Many of these substances that are found to be carcinogenic are used on our fruits and vegetables. That's why I say we are living in a carcinogenic environment. We are living it, drinking it, eating it.

The Coal-Tar Menace

Aspirin put on the back of a rabbit's ear will produce a cancer everytime - not just once in a while, but every time! Coal tar, a basic ingredient in that aspirin, is a carcinogen. Vitamin B, the kind put in enriched breads and cereals, also is a coal-tar product. This coal tar is the same product used to pave our highways. It may be high in Vitamin B, but because it is a coal-tar product it's not fit for human consumption. No wonder it causes an irritation in our body and proves difficult to digest. No wonder it has difficulty flowing with our natural flow and instead becomes an accumulative substance that eventually will irritate us.

Coal-tar products are in alkalizers sold in our markets, part of practically all the additives put in foods. That includes artificial flavorings and colors.

Look at the pesticides we spray on our foods! Some of us would like to wash off the spray on apples, but how can we do that when 75% of that spray is in the core!

Some people already have discovered how poorly man governs the air and water. Pine trees, one of the finest producers of ozone, a health-giving property in oxygen, can no longer grow in Los Angeles and other major cities. The smog kills them off. 95% of the natural water in this country is polluted today. We haven't gotten down and discovered what we are doing with food, probably the most healing factor next to the air and water.

Food just isn't that important to man yet. Man doesn't know what is foodless food as compared to chemically right - whole, pure, and natural food. There would be many red faces tonight if man recognized what is happening to the food that is meant to nourish his body. Look at all the processors of food, all the polishers and canners of food! Food production, most of it involved in changing food from its natural state, is one of the biggest industries man has.

Look At Our Soil!

As far as soil goes, we could be using more good soil that lies buried under cement and other man-created developments. We should be building homes in the hills and saving the valleys, but we built our homes in the valleys and now all we have left are hills with coarse soil. The topsoil is underneath those buildings. Government reports show me that once there was as much as four to six feet of topsoil 150 years ago. Today on an average we have less than eight inches of top soil in the United States. We "grow" from topsoil!

Half of our topsoil is in the Mississippi River. It's been blown away in dust storms - did you ever hear of the dust bowl? Millions of acres of topsoil have all gone out into the ocean. We don't know that every time we pull up a tree, we had better plant another one, because it's this vegetation that keeps soil from roaming.

We haven't learned how to take care of this soil.

Once in college they taught students about cover crops which are another way to protect the topsoil. We grow fava beans, a high source of nitrogen, on some of our land for this purpose. This removes the need for artificial nitrogen as provided by the oil industry.

When tomatoes are planted year after year, they shouldn't be planted in the same place. Every fourth or fifth year, the soil should be rested, not that the sun is left to bake on it, but another crop should be grown which will replenish the "used" elements in the soil again.

This is the way we did things, but times have changed. We just can't keep taking and not putting anything back.

Not long ago, cases of celery were picked up in Chula Vista, California, and had to be marked "Unfit for Human Consumption" because of the excessive sprays that had accumulated on these would-be foodstuffs. Celery, if left alone, would not cause us any trouble but that spray would!

Organic — Whole, Pure, Natural

We are getting less and less organic - whole, pure, natural - foods everyday. Who wants to grow foods for the market and have them bug ridden? If a grower can keep the bugs off and the food looks beautiful, whether it has the chemical balance or not, what difference does it make to the consumer. These diseased foods are sent right into the marketplace, and more and more of them are arriving each day.

We are not treating our bodies well in this day. This carcinogenic age is increasing, not decreasing. We are afraid to give up pesticides! If we gave them up, pestilence and famine would reign in our gardens. This would be one way Nature would show man how poorly he is governing his earth.

Back to Bugs — Some Conclusions

1. When food has the "tearing-down type" buglife actively working on it, it's a sign that this food isn't chemically balanced or

fit for human consumption anymore. The buglife in charge of decomposition should not be working until the healthy plant has passed on its seeds to the next generation.

2. Nature doesn't want seeds from a degenerative plant having offspring and bringing in a new generation. If the sick plant did propagate, Nature would not be refining the next generation. In a million years, the plant family should be stronger than it is today through natural selection.

3. Food that grows to perfection because it is chemically balanced has no germlife or parasite life working to tear it down until after it has ripened. This healthy plant needs no pesticides or sprays. Only the sick plant requires spraying to keep the parasites away. Those parasites should be there.

CHAPTER SIX

MORE ABOUT THIS AGE...

When we look at a tree we are looking at a finished product. A flower is finished. It comes up and lives its all and then it dies. It can't be anymore! But man can be something better all the time.

Man is never finished. We are so half-finished that it is unbelievable. We have so much more to do in expanding our consciousness, in expanding our minds, in being kinder to people, and developing a more natural knowledge of the soil and plantlife.

People today are going places which aren't too good for them. They may not know any better, but rather than feel sorry for them, bless them and instruct them. Those who are leaders have a terrific obligation to help those in need of instruction, and a terrific obligation to lift people.

Those who know no better, have a terrific obligation to listen and to reach for higher things in life so they can get out of their problems and troubles.

Somehow or another we don't follow the natural patterns that have been laid down for us. We have gotten off the path.

Listen to Some Doctors

"People in America are born with more nerve tension than in any other country in the world," one doctor tells me. After spending nine months in a very dark chamber, they are brought into bright light - no wonder Americans need so many glasses.

Imagine someone being in a room for nine months in total darkness and suddenly coming out into the light. What's more, the first thing they do is give the baby a good licking! They shock it the first minute it comes into this life. No wonder kids start out with such tensions. No wonder one out of every seven adults in American is taking tranquilizers.

Look at the animals. Kittens don't see the light the first day; why they're almost blind. There are places in Europe where they don't allow any child to see a full scale of light for about seven or eight days.

What a difference it would make if we brought them in with music! The New Age should have babies born to music.

A Look at Child Rearing

Look at our infant babes. First we take them away from mother's milk and put them on artificial foods and sweeteners. They are given too many starches which throws the chemistry balance off. The baby develops a larger pancreas to take care of the sweets. More quickly than we can imagine, he develops diabetes and so many of the other ills we must treat today.

Then he grows up a bit and we start our bribing. "I'll give you this candy to do this, and that candy to do that," Then we get to the place where we start taking candy away from people. They start crying, and some people just don't get over it.

Some patients come into my office as if something had been taken away from them. They just never grow up.

The Challenge Facing Youth

Youth today is trying to take on responsibility, and this is a terrific job. To take on responsibility without knowledge or direction leads to a tombstone. The greatest good we can do for a child is to influence him. To teach him can sometimes be difficult.

All we can do is put down in front of him all that is perfect, teach him what is right and wrong, and expect him to do the right thing as you were expected to do the right thing. Bring him to the nicer things in life, then we have done our job. A demonstration of what is right in life will sometimes convince a person that the right way is the way to go.

We have so few demonstrations today. We just tell a person and expect them to go ahead on blind faith. It is a case of the blind leading the blind. Visit a farm. You see compensation right away - right brings right. It's the man who has never been on a farm who says we need fertilizers for the bugs.

CHAPTER SEVEN

LOOK AGAIN, THE FIELDS ARE RIPE!

Doctoring in Parts

"Doctoring", too, must come of age.

Recently I attended a conference on healing in San Remo, Italy. More than 600 doctors from 34 countries were present, representing all segments of the healing arts.

One doctor says, "I operate and take an organ out. I no longer have to worry about that organ and the patient no longer has to worry about it anymore, either. It's finished."

A still wiser doctor says, "I'm a dentist. I found out that when I pull a tooth, there's a patient on the other end." And he continues, "I don't take a patient anymore unless he lives right so he will be making good teeth. He should be keeping his remaining teeth in good health through good nutrition."

When I spoke at this convention I stressed that we cannot diagnose people in parts. We have to look at them as a WHOLE being. We must take care of the totality in Nature. You can't think only of diet to get a person well, and just treating one organ in a person is wrong.

Further, just treating a disease in the body without treating the whole body is foolishness.

At that convention, it seemed that each doctor stressed his own specialty which was only part of the healing story. Healing should take in all of the arts - mechanical, chemical, spiritual. Just trying to cure man with breathing exercises or food isn't enough.

When a patient is treated, it is the doctor's obligation to expand the consciousness of that patient. When a patient enters with coffee and doughnut habits, is treated, and then leaves the office living on that same coffee and doughnut diet, he has not been elevated. He is not ready to make a better body. Nothing has been done to improve his life.

Stop Breaking Down

In order to follow the path Nature would approve of, and this must be done on physical, mental and spiritual levels, we must stop breaking down. Too many of us have lives filled with distraction and disturbance; we die frustrated. We seldom live a life of achievement the way we should. We don't live a life "overcoming and becoming;" most of us just exist.

Beauty Heals and Uplifts

I have my bad days just like any dog. I find that if I can spend time in my garden among my flowers the first thing in the morning, I find I talk differently later in the day. I think differently because I talk to flowers. Flowers are beautiful. When I come from them it's hard to turn to patients and not talk about beautiful things with them also. In my garden I am saturated with something lovely, and this affects my whole day.

I am beginning to understand what Luther Burbank meant when he said he would never hire a person who did not love flowers.

I asked a man what he does for a living. "Oh, I'm just a farmer," he says. I asked a lady what she did and she said, "Oh, I'm just a housewife."

We must find beauty in all we do, even though we must do some things we don't like. Unless we change our attitude, we are not going to get the good things out of life.

Ruts

We get in a rut; get lazy. We say, "Oh, I can't do this, and I can't do that."

Get over this attitude! There is no such thing as an impossibility. Too many men are sick because they believe they are supposed to be sick.

I visited a little lady in Anaheim once. Here she was lying in bed. "What seems to be the trouble?" I asked.

"I've got heart trouble," she said, "and I've been in bed three years."

"When did you see your doctor last?" I askèd, and she said, "Three years ago." "You don't mean it! Have you been in bed for three years?"

"Well, the doctor told me I was incurable and I just had to rest, so I have been resting for the last three years."

If she had lived three years resting in bed, there was a possibility to strengthen her heart through proper supplements and foods.

I convinced her it was possible and she eventually got out of bed and lived sixteen years.

Don't believe every impossibility you hear.

A Load of Sharks. . .

A cargo of live fish was coming from Norway to New York. On the first boat load, all the fish were dead when they arrived. On the next boat over, a little shark was put in the hull. When the second boat arrived, all the fish were alive. Those fish were moved! The exercise kept them alive.

Sometimes we need to move too. We need a little shark in our lives - something to move us along. Sometimes it takes a little disorder and chaos before we do a little thinking. I hope that most people don't have to wait too long to do something good for themselves.

. . .And A Frog in the Road

Once a frog was caught in a rut in the road. All his friends came by and tried to get him out of that rut. No one could help him, so they left him there.

Half an hour later, this frog was hopping down the road, chirping and singing. All his friends gathered round full of inquiry.

"My goodness," they said, "How did you ever get out of that rut in the road? It looked like an impossibility."

"We couldn't help you. Nothing could be done!"

And the frog said, "Well, I want to tell you. A truck came along, and I had to get out!"

This also is the way some people move! They see a truck coming. Then they really move! Sickness can make us act this way too.

A Friendly Hatpin

We have to come to the place where we have an AWAKENING. That is all I can hope to do for anyone. Many patients feel that as a doctor I can do this or that, but I really can't do very much. All I can do is give you a little hatpin. Push you along, awaken you, get you started on a new path. I can give you the ball, but you have to carry it.

I know Nature works, but I can't work it for you. All I can do is share it with you. I've been sick myself. I know what it is to be down. I know what it is to have something and then lose it, to have your happiness taken from you through loss of a loved one. I also know what it is to start over in life with a year old babe under your wing.

I know too that the ball that hits the ground hardest, always bounces the highest!

I know that we need the experience of living with sickness and living in this present age. The fool can't understand what's in the wise man's head. The wise man understands what's in the fool's head because he was once a fool himself.

That wisdom you need is coming to you. I don't wish you too many more bad or sad experiences, but I do know you're going to come out of them, because there is only one way to go, and that is up. There is no experience that you will have that you won't be better for having, no matter how bad you consider it.

In This Survival Age

Things may be taken from you, depreciation may come to you, and losses may strike you. Forget the idea of losses and start attaching yourself to all that is good. There is never a negative without a positive charge right next to it. There isn't any heat in the world without some cold. There isn't any sunset that ever comes, that a sunrise doesn't follow.

Right there where you look, you will find the good, for the good thing has been said, "Look again my friend, the fields are ripe and ready to be harvested!"

Part II

Chlorophyll Magic
from
Living Plant Life

The Story of Nature's Greatest Healer
and a Nutritionist's Time-Tested Remedies

PREFACE
To Four Survival Foods

In this day and age, our environment has become chemically sick and everyone is concerned about cleaning up the air and the water. We must also clean up and provide a more perfect food for our body's use. Clean food for the body is just as important as clean air and water. We have gone so long on degenerated foods that doctors are making a living on OUR living.

At a recent lecture, I asked the audience how many knew that cigarette smoking was bad for them, how many knew alcohol was bad for them, how many knew that coffee was bad for them? Many hands went up. When I asked them what was good for them, however, few could tell me. It is about time we learned what is good for us. It is about time we become wise enough to select what is good for our body economy!

Our Survival Depends on our Daily Food Habits

When we consider the future of mankind, there are four crucial survival foods which should be in everyone's pantry. These four survival foods are the beginning of a program which can take you into the whole realm of what is good for you. There was a day when everyone talked of wonder drugs. Here are wonder foods for humanity. If you are pushed against a wall and have to survive a disaster or a disastrous disease, they could help you. However, you should realize that survival today isn't a matter of waiting for an earthquake or a disease, but rather is a question of our daily living habits and what foods we put into our mouths. These foods which I consider of highest survival value are seeds and nuts, sprouts, berries, and chlorophyll. These are foods you must learn to master.

Master these Crucial Survival Foods:
Nuts, Seeds, Sprouts, Berries, Chlorophyll

Learn how to make milk substitute drinks out of the nuts and seeds, the glorious ways to grow the sprouts and use them, how to create berry drinks, and how to unleash the power of chlorophyll in such ways that even children will look forward to receiving their daily greens.

These Crucial Foods are Part of Our Magic Survival Kit

In previous booklets and talks, I have discussed the values of nuts and seeds, sprouts, and berries. I hope to offer some advanced ideas about these foods in future lessons. In this book, we are going to take up chlorophyll which is a subject I have not treated previously. There are many people who would like to clean up their body and do it quickly. There are many people who want to do it without any hemming and hawing around and they don't care whether the remedy tastes good or not.

We are giving these people the cleanest, purest, and most natural adjunct to their diet, a liquid sunshine which is the most health-producing fluid they can put in their body.

Health Magic Through Chlorophyll

For over forty-five years, I have dealt with chlorophyll foods. I have produced them, prepared them, and created many remedies using them for patients. Throughout this book, I stress that the potential of chlorophyll is a mighty weapon, a weapon you can have in your diet for maintaining or retrieving health, and this potential is yours if you live right and follow a healthy living regime. If you continue on a coffee and doughnuts diet, there will be little magic for you.

I hope that going into detail about one of Nature's greatest healers—chlorophyll—will give you some insights and start you reaching for some of the better things in life and the health you are trying to attain. Let us begin our work.

AUTHOR'S NOTE: While many may construe the remedies in this book as home remedies which may be used on their own accord, in any serious condition it would be well to consult your doctor. I believe your doctor would be very interested in the nutritional side of chlorophyll and in reading this book.

I present this material from some of my studies. The remedies have been used with patients over a period of years, and I would like to suggest that here is common knowledge of the future, and remedies which many doctors will be recommending. Many of the ideas and remedies are not on the "market" or thought of at this present time.

Although the benefits of chlorophyll are available to most people, I would like to caution that in some rare cases allergic reactions have been reported. It is best to consult your doctor and discontinue use of greens if negative results occur.

CHAPTER EIGHT

INTRODUCTION TO CHLOROPHYLL

"Finally, he listened to them as regards this matter, and to put them to the test for ten days. And at the end of ten days, their countenances appeared better and healthier in flesh than all the children who were eating the delicacies of the King. So the guardian kept on taking away their delicacies and their drinking wine and giving them vegetables."

Book of Daniel, Verse 14:16

A Few Words from Daniel. . .

The Bible tells us how Daniel the Prophet was asked to interpret the King's dreams. He refused the King's wine and meat because he did not consider them foods for a perfect diet. Daniel, being a wise man who took care of his body, lived on the Pulse (greens of the field) and Water for ten days. He ate parsley, turnip and beet greens, and watercress, and He Came Back Fairer Than All of the King's Youth.

I don't know how much fairer the average person would like to be. When Daniel went to see the King, the monarch could not remember his own dream, and Daniel recalled the dream without any hint of its content and then interpreted it! Daniel's mind was able to do this. He was truly a King over his body and mind.

Why did he have such faith in the greens of the field?

Minerals: From Dust to Food

When we go to a tree, we find that it has calcium, silicon, sodium, iodine, and many other chemical elements. This tree and all plant life have all that was the "Dust of the Earth." All of the tree has been made and is nourished by its blood which is chlorophyll.

Chlorophyll is the green pigment that plants use to carry out the process of photosynthesis. Its function is to absorb the light energy used in photosynthesis for the reduction of carbon dioxide to sugars and other plant materials. Heat, water, and sunshine are also important parts of the process.

Without sunshine things don't turn green. Notice that when you are sprouting seeds they are a real pale green, but put them out in the sunshine and they become very, very green.

Lois Mattox Miller, writing in the *Science News Letter* for March 15, 1941 makes some beautiful statements about chlorophyll in her often-quoted article:

"A ray of sunlight strikes the green leaf and instantly the miracle is wrought. Within the plant, molecules of water and carbon dioxide are torn apart—a feat which the chemist can accomplish only with great difficulty and expense. First there are only lifeless gas and water; then, presto! These elements are transformed into living tissue and useful energy. Oxygen is released from the plant to revitalize the air we breathe. Units of energy, in sugars and other carbohydrates, are speedily manufactured and stored in the living plant.

"Out of the process stems much of what we know as life and growth. Man consumes the energy as food—both in vegetable and the flesh of herbivorous animals. He uses it in the form of coal, oil, and gas—green vegetation locked up in the earth for ages."

The life-given quality of chlorophyll is emphasized by the following quote which appeared in the *Organic Consumer Report* compiled at Eden Ranch, Topanga, California:

"Photosynthesis makes the food we eat and the air we breathe. Back in 1771, Joseph Priestley, an English clergyman, found that a mouse in an airtight glass bowl soon began to suffocate, but when a sprig of fresh mint leaves was added, the mouse revived. It was later explained that the oxygen under the glass was being renewed when sunlight played on the green leaves, and soon the mystery of photosynthesis became a challenge to scientists."

That mystery still goes on.

Photosynthesis: Where God Begins

I have really never heard a good complete explanation of photosynthesis; in fact, science claims there isn't a good explanation of it. This is where man ends and God begins. This is where God begins to make the materials for man out of the "Dust of the Earth."

If God did make the soil, make the earth, make the chemicals, they certainly aren't any good for man's use until the energies from the soil and sunshine have been transformed into plant life. Right there in photosynthesis is where God has more to do with helping mankind than any place else.

Personally, I feel God and Nature are practically the same thing. We talk about a natural force, we talk about a divine energy—it's all God.

All Animals Need Greens

Many times I have had patients bring their sick dogs and cats into my office for treatment. We find we even civilize our pets to the point where they haven't health either. If left to ferret out foods for themselves they definitely would be having more greens in their diet, but in the city, how can they get them? In New York City they have a 30-foot square of grass with a sign which reads "Please do not step on this grass. Keep it alive. It's the only grass in a ten mile radius."

They claim I can take care of human beings, so why can't I cure their dear pets?

The first thing I do is make a fresh grass juice or get some concentrated liquid chlorophyll for the animal and squirt this down his throat with an eye-dropper. Many times they will drink the juice from their water dish. Then we have them swallow garlic capsules. These are also dropped directly into the throat so there will be no chewing and therefore no regurgitation.

Next we begin a rebuilding job with fresh greens. The pets, especially the ones which have been living on canned food (which has too much cereal in it), respond to this natural treatment. Observe animals. They go to the fields for these nice grasses when they are not feeling well.

Grasses are Fundamental Foods

Whenever anyone tells you to "go to grass," GO!

I believe grass is probably the most fundamental of all foods we have. Herbs started from grass and most vegetables came from herbs. Then finally, bigger growth came from the vegetation and developed into our trees and bushes.

Grass doesn't have to be planted. Take any little piece of ground, clear it, water it, and in four or five days grass starts coming up. No encouragement is needed. All it needs is the proper environment—water, sunshine and the soil.

Although we would have a hard time getting all the nutrients we needed just from the common grass we see around us, many of the green brothers of everyday grasses do play a crucial role in good health.

"You still don't welcome the thought of eating grass?" writes Ben Harris in his book *Eat The Weeds,* "Then, pray tell, why at breakfast do you eat farina or oatmeal or wheatgerm grains (or even toast, for that matter)? These cereals are members of the grass family and others more familiar to us (out of 1000 assorted relatives) include bamboo, sweet corn, sugar cane, barley, oat and rice."

Says Harris, "It has been estimated that 12 pounds of powdered grass contain more vitamins than 340 pounds of vegetables and fruits - more vegetables and fruits than the average person can eat in a year."

> *"My tongue, every atom of my blood*
> *formed from this soil, this air. . ."*
> *Walt Whitman*

When the Good Book says that it is in the blood that we have our life and our being, stop and think of the chlorophyll we get from greens and the type of body it builds.

Chlorophyll helps build all of Nature. For a tree, the leaves and bark, sap, limbs, seeds, juice—and all of these parts of the whole have the same chemical elements which are found in the human body. Chlorophyll can be used to help improve almost every disease in the human body where these chemicals are short.

Greens: Basis for our Meals

Looking to Nature, we find so many beautiful fruits and glowing greens there. These foods are cooling and rejuvenating and should be the basis for our meals. We begin with a salad, and think of the green leaf lettuces, the Romaine, the Bibb, the Red Tip, for these have 100 times as much iron as head lettuce, a bleached-out vegetable low in chlorophyll which is commercially grown for mass harvest and sale. Consider also that head lettuce contains an opium-like substance which slows digestion.

Green, A Healing Color

It is no haphazard occurrence that the color green is so prevalent in Nature. Green acts on the nervous system as a sedative. We find it is helpful in insomnia, exhaustion and nervous irritability. The color green can lower the blood pressure, dilate the capillaries, produce a sensation of tranquility. Neuralgia, headaches, and nervous conditions often respond to its use.

I believe green is a prime healing color. The pine tree has been called the "sacred" tree. In Central America, the natives place pine needles around their idols believing this will be a means of bringing everlasting life to the natives. Painting with green colors has helped many mental patients to have an improved outlook and mental attitude. This is because of the cooling effect which slows down the fast vibratory force of an anxious and apprehensive person.

Green is a calming color, the restful color. It is suggestive of life and immortality, the emblem of youth and happiness and prosperity. It connotes a sense of well-being, good health, and abundance. It counteracts the brightness of the sun, excites the eye less than any other color, including black, and induces repose.

Chlorophyll: Food for Tissue Repair

When we take this green into our body, the results have been short of miraculous, for chlorophyll is an all-around food beneficial for tissue repair. Anyone who takes liquid chlorophyll is helping to neutralize some of the sprays and abnormal materials from artificial fertilizers that get in many of the foods we buy. Liquid chlorophyll builds up the health so much that the body can take care of these abnormal deposits. Further, we find that chlorophyll is very easy to absorb and assimilate. It gets into the bloodstream with the least amount of digestion which saves a lot of the vital energy that usually goes into taking care of the average food's digestion.

I am sure that in the beginning, greens were our main source of food, just as surely as the Garden of Eden was a green garden.

A plant is alive because of chlorophyll. A dried-up brown leaf has no appeal. When it is said that God has given the seeds of the field and the herbs of the field, remember that He has also given us the GREENS of the field. The greens, so powerful that some animals live on them entirely, are also for man's use.

How I Got Started With Greens

In my nutrition work I have had many failures, but also many good results. One of my nicest cases was a young lady who came in with 13 leg ulcers running yellow-green pus for three years. She had been in many institutions around the country and finally came to me for diet advice. Her case had been diagnosed as pellagra which is caused by a lack of calcium. The materials necessary to control calcium were not in her body.

This little girl started me on my chlorophyll search. When she came to me, I was just getting into diet work and still had a lot to learn. I asked myself where I could get a control of calcium in the body. realizing that calcium is the first chemical element to consider in healing.

Greens Control Calcium

I read a book called *The Newer Knowledge of Nutrition,* by McCollum and Simmonds which made me conclude that the long-lived Hunzas were controlling the calcium in their body with greens. They had no pellagra or rickets (where bones are brittle and break) and died with every tooth in their heads and with wonderfully healthy calcium in their bodies.

I also read a government report that stated that only 16% of meals in the United States consisted of vegetables. More than 60% of the food eaten by the Hunza were sunshine vegetation. If they had such wonderful calcium control, it must have come from their sunshine vegetation.

BEFORE

AFTER

These photos, reprinted here from an earlier publication, illustrate one of Dr. Jensen's first experiences using the greens of the field in natural healing. This woman had a complete recovery from 13 leg ulcers within three weeks.

I learned more by studying various experiments. In 1915, when a lack of calcium showed up in the health of the South, Dr. Goldstein was given the nutritional award for suggesting if all the people would eat more turnip greens, they could alleviate the disease pellagra which came from a lack of calcium. The common practice, then, was to plant cotton all the way up to the houses which left little room for a garden.

Chlorophyll: A Sunshine Juice

At that moment I recognized chlorophyll as a sunshine juice, a sunshine liquid for the body. Chlorophyll, I realized, is only produced when sunshine is present. When we want to control calcium, we find that greens, a sunshine vegetable, are to be used. Chlorophyll has a more direct control on calcium than anything else. If I wanted to help her keep calcium in her body, I knew the answer would be giving her sun-grown fruits and vegetables.

Making a Vegetable Top Liquor

I took nine tops of vegetables—celery greens, beet greens, turnip greens, and watercress and many other leaves with large arteries. I chopped them up real fine, left them for an hour or two in distilled water, and then squeezed them through cheesecloth for her to drink 3-4 quarts a day.

In three weeks' time her ulcers disappeared. To see how a healing could occur so wonderfully before our very eyes really moved me and gave me the dedication to pursue this food work.

Chlorophyll Makes Quick Cell Changes

I also recognized many other facets of chlorophyll through my studies. For instance, it had an enormous effect on egg yolks. When chickens were fed greens, they would have a nice dark orange egg yolk. To see this change! It all came in the matter of one or two days, showing that the effect on the body was taking place almost immediately. To think that the whole body was making a change.

That egg had to be made from that whole body! An egg is no better than the chicken itself.

Chemical Shortages Cause Diseases

Since then, I am always encouraged when I find medical doctors or others in the drugless healing fields that recognize the role of nutrition on tissue growth. I think nutrition is the fundamental healing art and should be considered first. Our body health has a direct relation to the nutrients you can get from fruits and vegetables, seeds, herbs, and so forth.

Dr. Tom Spies, M.D. made these statements before the annual meeting of the American Medical Association in 1957:

"All diseases are caused by chemicals. All diseases can be cured by chemicals. All the chemicals used by the body except for the oxygen which we breathe and the water which we drink are taken through food. If we only knew enough, all diseases could be prevented and could be cured through proper nutrition."

He continued by saying, "As tissues become damaged, they lack the chemicals of good nutrition. They tend to become old. They lack what I call 'tissue integrity.' There are people of forty whose brains and arteries are senile. If we can help the tissues repair themselves by correcting nutritional deficiencies, we can make old age wait."

Dr. Spies was given an award by the medical profession in 1957 for great contributions to the healing art. I hope many more like him fill the ranks in the medical field.

Out of Vancouver, they have found that one of the greatest dissolvers for chloresterol is found in sprouts. Dr. Evers in Germany is doing things with multiple-sclerosis that are "out of this world" and he is doing it with greens from sprouts and other chlorophyll sources.

A Remedy For Sores

Some other wonderful remedies developed from my association with chlorophyll. I used a chlorophyll salve for external ulcers and other skin disorders. (Mix concentrated liquid chlorophyll with lanolin and add garlic oil from one capsule. The chlorophyll will be about 50% of the mixture. This salve will also promote healing for acne, pimples, and boils).

Here in chlorophyll was one of Nature's greatest healers both for conditions inside and outside the body.

Because it is so prevalent in Nature it reminds me of the Armenian saying, "God created the cure before the disease." I mention many times and repeat, "When you are green inside, you are clean inside." Using chlorophyll is making food your medicine.

My Visit With The Hunzas

I wanted to visit the Hunzas for many years after I had learned about their use of sunshine vegetation, but it wasn't until 1971 that I had the opportunity to travel there.

When I arrived in the valley, these people were using as much as three or four teaspoons of white sugar to every cup of tea they were drinking. The use of white sugar was a recent habit they acquired from the outside world.

Both my wife and I went out and picked alfalfa and some nice wild peppermint, and I suggested to the Mir that he use such a green alfalfa-peppermint tea.

Dr. Jensen with the Mir of Hunza

"You know," he said, "you can even drink this without sugar. It has such a wonderful flavor."

Well, to have a cup of tea that was alkaline-forming and put it in place of an acid-forming cup of tea, as the white sugar provided, was probably the greatest deed I could have done for the Hunzas from a nutritional standpoint. Maybe, in some way, it was giving thanks to them for the knowledge of the greens which I had learned about in their diet years before.

Chapter One Summary

1. *The role of greens (the pulse of the field) in producing a healthy way of living has been known since ancient times. Wise men select the food they eat. Greens are foods wise men have chosen.*

2. *Although much about the process of photosynthesis is unanswered, man knows that in this process inorganic earth elements - "The Dust of the Earth" - are transformed into bio-chemical minerals. Where man can not benefit nutritionally from eating these earth elements directly from the soil, the plant provides a go-between. These vital transformed elements are the basis of man's nutrition.*

3. *Chlorophyll, the green pigment in plants, plays a vital role in the process of photosynthesis. Because of chlorophyll the power of the sun is captured by the leaf. This sunshine value is available to man when he eats green vegetables and their brothers (herbs, grasses - grains, etc.). Many of the benefits man gets from the sunshine he can get from greens.*

4. *Green plants, when in the presence of the sunshine, create life-giving oxygen for man's use.*

5. *Sunshine-grown vegetation is a calcium control in man's body. Diseases caused by an inability to hold calcium in the body, such as pellagra, are benefited by greens in the daily eating regimen.*

6. *Where the bark(skin) or fruit of a tree may contain only a PART of all the chemical properties of the tree, chlorophyll, the bloodstream of the tree, is the WHOLE of that tree. It carries every life-giving mineral the tree has. Chiorophyll builds all of Nature.*

7. *The chemicals of good nutrition are in greens.*

8. *Tom Spies, M.D., notes that all diseases are caused by chemical shortages. Because chlorophyll is so much a WHOLE food in Nature, it can help man wherever he comes up chemically short of vital body nutrients. It will help any nutritional shortage.*

9. *Chlorophyll is an all-around food beneficial for tissue repair and building. A chicken fed greens has shown dramatic changes in the color of the egg yolk within a few days, indicating all body cells were affected.*

10. *Chlorophyll is one of Nature's greatest healers for conditions inside and outside the body.*

11. *Greens should be the basis of our meals. Much of Nature's garden is green. Much of man's food must also be green if he is to remain in balance with the Natural environment and Laws of Nature.*

12. *Teas made from herbs are health building. Because they are aromatic and flavorful, they can help us cut down on "tasty" health-damaging white sugar as a tea-time sweetener.*

CHAPTER NINE

CHLOROPHYLL CONSIDERATIONS

Sardines and Seeds: "I can't believe
I ate the Whole Thing!"

In 1973 I visited Charlie Smith, the oldest man in the United States. One of the few foods he eats is sardines. Sardines. You eat the bones, nervous system, stomach, glandular materials. The "whole thing" is in that sardine and one of the Laws of Nature is to eat the whole food. All the canners did was take a few of the scales off, and they even left some of them on.

It is also a valuable food because it has all the trace minerals, and all the other ocean chemical elements. The ocean has the greatest balance of chemicals, and the sardine, as a small fish, has been given every element in a concentrated form.

Some Natural Food Notes

I regard seeds highly because they also provide "every element in concentrated form." The tendency today is to get away from producing fruits and vegetables with seeds and instead produce larger and larger specimens. This is a step in the wrong direction. The smaller the fruit, yet the one containing all the elements, is the better way to go. Berries are valuable because they have so many seeds and because they are close to their original form. Seeds are also valuable

---

SURVIVAL LAW NO. 4

Man was meant to eat WHOLE foods. Whole foods come from Nature's garden as Nature intended them to. They are not refined, processed, toasted, roasted, oiled, or greased. Nothing is added or subtracted. Man comes up chemically short, i.e. unbalanced - sick, when he fails to eat the WHOLENESS which Nature provided. This is the Law.

---

because they pick up these trace minerals. When man creates hybrids, such as these seedless fruits and vegetables, he is interfering with Nature. Leave pollination to the bees.

Trace Minerals

We find that chlorophyll also compares in value to the seafood from the ocean. AS SEAFOOD HAS THE GREATEST AMOUNT OF TRACE MINERALS FROM THE SEA, SO CHLOROPHYLL HAS THE GREATEST AMOUNT FROM THE LAND.

Plant Water is Our Best Water

All the trace minerals are here, and given to you in the purest fluid possible—a distilled water. When the plant brings water up from the ground, this water has to go through the root structure and stem structure, from one cell to another, and when these liquids get to the leaves, they have been refined and distilled many times.

This process picks up many coarse mineral elements also, but most importantly, as the distilled water goes through the fibrous material of the plant, much of the coarser materials are filtered out and left behind. The fluid takes up many of the soluble and absorptive trace minerals which will be the easiest chemicals for our body to accept. It is not surprising that our bodies have a very special need and use for these fearfully-made particles which are so broken down that they are almost cosmic in nature and no longer discernable from a physical standpoint.

The Lymph Stream

In most cases, the regular bloodstream feeds the tissues of the body, removing waste materials and bringing in materials for tissue growth. Sometimes the regular bloodstream actually deposits new materials. Sometimes it is just the influence of these chemicals that does so much good for all the cell growth.

However, there are delicate and intricate areas of the body where the regular bloodstream cannot enter because its cells are too large to get in and feed these areas. For this reason, we are equipped with another "bloodstream" called the lymph, which works independent of the regular stream which carries the red corpuscles. This lymph stream feeds and carries waste from these delicate areas, such as the cartilage in the knee and the lens of the eye, which require these trace minerals and this finer chemical material.

Feeding the Eyes Nutrients - Osmosis

The lens of the eye provides us a good illustration of the role these trace minerals can play. Here we have an area as big as a thumbnail yet more than 3000 layers thick, one on top of another like the skin of an onion. In order to replace dying cells in this lens, (and all live tissue must be replaced with new tissue at the end of a specific cell-life span), materials for the development, rejuvenation, and the regrowth of new tissue must be carried in.

Because there are no arteries in the lens to carry in the materials for cell development, the lymph must come to the rescue. Through a process called osmosis, which allows for seepage of these fine trace minerals, the new materials are exchanged. Because the lymph does not rely on a pump like the heart to get itself around the body, it requires exercise to move it. The particles float through the layers bringing materials for rejuvenation.

Chemical Imbalance Causes Cataracts

In the case of a cataract, an improper chemical balance exists between the elements necessary to balance calcium, namely sodium. The hard calcium has come out of solution and deposits itself in the various layers of the lens. The calcium becomes milky in color and gradually you can't see through it.

By dissolving through osmosis, there is an exchange of coarse materials for materials which are finer. A balance is again created. Sodium, the youth element which keeps everything soft and limber in the body, has an opportunity to get in there. Calcium, when it comes out of solution, makes these tissues hard wherever it deposits itself. A cataract is the same as an arthritic formation.

The body can, however, with the chemical elements and the help of these balancing trace minerals, rebuild, replace, and exchange, and this is where the new tissue comes from.

Chlorophyll Holds Trace Minerals

Chlorophyll, because it contains so many of the "biochemical" trace minerals, meaning they are fit for human consumption, is one of the most wonderful foods we can have. I recommend sprouts and green juices heartily because they have the trace minerals. Here is a food that can get into these delicate areas! Great improvements in health can be made with it.

Because these trace minerals are obviously so important to good health, we should make sure we are eating foods that contain them. Again, few processed foods do.

Chlorophyll: A Relative by "Blood"

I don't think anything will enter the bloodstream faster than chlorophyll. Amazingly, from a chemical standpoint it has almost the exact components of blood. It was Dr. Richard Willstatter, the German chemist, who in 1913 determined that the chlorophyll molecule closely resembled hemoglobin, the red substance in the blood that carries oxygen to the cells. He determined that hemoglobin was composed of carbon, hydrogen, oxygen, and nitrogen atoms grouped around a single atom of iron. Chlorophyll has the same elements; however, the center is magnesium, not iron.

I believe someday man will take this knowledge and discover ways to use the green fluid to build the bloodstream rather than depending so much on the liver products to do so.

Transfusions may be necessary when someone needs blood in a hurry, but it's encouraging that we can all build and replenish our own bloodstream through proper nutrition with foods like chlorophyll.

Chlorophyll is Absorbed Almost Immediately

One of the finest foods for building the red blood count in the body is the chlorophyll in concentrated or vegetable juice form. I have had experiments where we have taken people, put them in a chlorophyll bath, and raised their blood count. I have given them vegetable juice, mostly the green juices, and we have been able to build a blood count.

These patients took a half-hour soak in this warm chlorophyll bath. We took the liquid juice from alfalfa and put it with the whole liquefied alfalfa in a warm bath, one-third full. A daily bath over three weeks raised the blood count 400,000.

I can imagine some smart person coming along and inventing a chlorophyll-soaked innersole to be worn in the shoes. To me, this wouldn't seem impossible at all, and I feel one of these days it can be done and will be done. Today, I am able to control an underactive thyroid by painting iodine on the soles of the feet—the largest pores in the body exist here.

Such an application allows the patient to absorb the iodine the body needs throughout the night. The idea of using chlorophyll on the soles of the feet is not far off.

Body Builder/Energy Saver

The ability to absorb chlorophyll is due to these small infinite particles that can be found in it. When you consider how most foods must be broken down by digestion, sometimes ranging from 4-6 hours, the ability of chlorophyll to enter the bloodstream almost immediately seems amazing. Here is a food with massive benefits, yet one which requires little loss of the body energy which is required for digestion, assimilation, and absorption.

Something to Think About

1. *Trace minerals are finely wrought chemical elements which travel though the fibrous material of the plant. When man eats the tender leaves of the plant, he is taking into his body the most infinitely soluble particles capable of feeding the most delicate and intricate parts of his body. Recent experiments indicate that long life and high I.Q. are related to trace mineral intake in body tissues.*
2. *Delicate and intricate areas of the human body exist, such as the lens in the eye, which cannot be fed by the coarser element-carrying bloodstream. The lymph stream, working independent of the bloodstream, must feed these areas with trace minerals and other finely composed minerals.*
3. *As seafood has the greatest amount of trace minerals from the ocean, so chlorophyll has the greatest amount from the land. The greenest chlorophyll plants have the greatest amount of trace minerals of the plants produced on the earth. Refined foods or man-handled foods seldom contain trace minerals in natural form.*
4. *The finest water obtainable is the distilled water as it comes from plants. We can get this chlorophyll-rich water when we drink vegetable juices or eat salads, etc.*

5. *From a chemical standpoint, chlorophyll has almost the exact components of human blood. It is an excellent red cell builder for man. It is the greatest food we have to bring someone out of anemia.*

Tips for New Age Earthmen

1. *The smaller the fruit, yet the one containing all the elements, is best for man's health. Less body energy is lost trying to extract the same nutrients from a tiny seed than a 25-pound cucumber.*
2. *Seeds provide every element in concentrated form.*
3. *Fruits produced without seeds are unfit for man. Seeds are crucial for glandular balance. They carry an electrical and radiation of benefit to the whole body.*
4. *Berries and many of our wild greens are original foods. They have not been "watered" down as many of our hybrid vegetables today. These original foods have an untouched WHOLENESS as Nature intended.*

SURVIVAL LAW NO. 5

Chlorophyll is a survival food because it feeds us dynamic nutrients yet our body gives up little energy to get these nutrients. The New Age man must consider eating foods which give us more energy than we have to put out to get the food digested and to our cells. A tired body cannot afford to give up unnecessary energy if it is going to repair and rebuild, get rid of the old, and make a new body.

Overeating, or eating unusual combinations of foods, demands heavy energy loss.

In this day when there are so many activities causing energy drains, we have to get the fastest and best energy - producing foods - chlorophyll foods.

Nothing is Foreign: Parts Relate to Whole

Look round the World; behold the chain of Love,
Combining all below and all above
See plastic* Nature working to this end,
The single atoms each to other tend,
Attract, attracted to, the next in place
Formed and impelled its neighbor to embrace.
See matter next, with various life endued.
Press to one centre still, the general Good.
See dying vegetables life sustain,
See life dissolving vegetate again.
(By turns we catch the vital breath, and die)
Like bubbles on the sea of matter born.
They rise, they break, and to that sea return.
Nothing is foreign; parts relate to whole;
One all-extending, all preserving soul
Connects each being, greatest with the least;
Made Beast in aid of Man, and Man of Beast;
All served, all serving, nothing stands alone;
The chain holds on, and where it ends, unknown.

Alexander Pope
(1688 - 1744)
Essay on Man III

*"plastic" means molding here;
"endued" means endured with

CHAPTER TEN

SUNSHINE

In years past calcium was controlled by putting patients under ultra-violet light, an important part of sunshine. An average cost per treatment was five dollars. Today, the sunlamp treatments aren't even given anymore despite ultra-violet's ability to control calcium in the body. Luckily, there are other more natural means of controlling calcium, and greens are one of these methods. Through greens, we can get all the benefits of the sunshine directly to the bloodstream instead of taking it in by lamp treatments.

The Sun Helps Us Grow

Although the sun plays a dramatic role in the life of all plants and animals, the average person doesn't recognize its value. Plant life begins in the darkness of the earth. The root goes down, while the life principle goes up, reaching for the sun. As soon as the plant life gets above the earth's surface, it turns green: Photosynthesis is taking place from the activity of the sun.

55

We know that without the sun WE cannot grow. This sunshine element is in the chlorophyll, and the chlorophyll gives us the sunshine effect. All life's dependence on the sun can be shown in a simple experiment. Cover a lawn with a canvas cloth for a week and the grass underneath will die.

City People are Denied The Sun

Dr. John Ott of the Light Research Department of Sarasota, Florida, has graphed and shown how 85% of the ultra-violet light from the sunshine is filtered out in the smogged cities. He has also shown that 85% of the ultra-violet rays are filtered out when we wear glasses. The greatest amount of good from the light value comes through the pupil of the eye which is covered when glasses are worn. The people who wear glasses should recognize that here in chlorophyll is a great way to get that sunshine element to the bloodstream which they miss by wearing glasses.

That our body controls light and color intake through the pupil of the eye has been shown in an experiment where one eye of a chameleon is covered with a patch. Only the part of the chameleon's body on the side where the eye has been uncovered will still respond to changes in the environment. The side with the patch on the eye does not make any changes.

Cities Lack Oxygen Too

The sunshine situation in the cities is so bad that people are dying of a lack of oxygen as well as suffering from a shortage of natural ultra-violet rays. When the smog comes heavily over Los Angeles, the hospitals have an increase in asthma cases the day after.

These smog-ridden cities also have a high percentage of babies who die one or two weeks after birth. The doctors would be helping educate the people to this dire problem if they would be more direct on the death certificates and simply write, "died due to the smog."

We Are No Longer Sunshine Children

Going down to the South Seas, the natives say, "What are you doing, building a house to get away from the outside? Then you put windows in to allow some of the outside in. You people are crazy!"

On top of all this, we put in window glass which filters out 85% of the ultra-violet when the sunshine comes through the glass.

Today, we are no longer sunshine children. We are apartment house children. The typical office worker sits in a little room with no windows and under fluorescent light (which destroys vitamin A), and sits there all day long. Anyone in the city should especially think of greens as a means of getting sunshine to the body.

Greens Bring the Sun to Us

Dr. George Crile, M.D., said once at a medical convention that in the future man will try to get the sunshine element from the plant to rejuvenate the human body. I look forward to more thinking along these lines.

There is a lot of chlorophyll in any food touched by the sun. Chlorophyll is not only in the green leaves (although greatest here) but also in most of the fruits of the trees. It is in the distilled water in grapefruit, oranges, peaches, tomatoes, etc. It's even in pecans. Every part of the tree depends on chlorophyll. Chlorophyll is the life force for the plant; it is also a life force for us.

Sun-Touched Foods

Those foods which remain in the sun the longest are the finest in their chemical structure for the human body. I do not believe in mushrooms and radishes because they are very quick growing. I also shy away from recommending citrus fruits for my patients because tree-ripened sodium-rich fruits are a rarity. It is essential that citrus fruits remain on the tree for a long time until they are completely ripe. Otherwise their reaction is to stir up body acids. In cases of toxic-laden acid bodies, as in arthritis, and/or where the elimination organs are functioning poorly, these acids get stirred up faster than the body can carry them off. One of the reasons I would rather use the vegetable kingdom than the fruit kingdom is because the vegetables carry off these acids. Chlorophyll just doesn't stir up the acids!

Citrus fruits may be high in Vitamin C, but they lack the chemical-mineral structure of greens. Chlorophyll will build the whole body but you can't live on citrus fruits.

The fruit of a tree is less a "whole" in relation to the total plant than chlorophyll is. If it was the citrus seed, it might be a different story. Chlorophyll is the equivalent of the bloodstream of the human being. It feeds the "whole" plant life.

Chapter Three Summary

1. *Sunshine controls calcium in the body. Greens also control calcium in the body.*

2. *People in smogged cities are being denied the benefits of the sun today. 85% of health-giving ultra-violet light is filtered out in the city by smog, window, and eye glasses.*

3. *Without sun we cannot grow. Dr. George Crile, M.D. predicts man will get the sunshine benefits from the plant to rejuvenate the human body. Through greens we can get the sunshine benefits to our bloodstream.*

4. *When citrus fruit is not allowed to mature in the sun, the unripe fruit contains acids which cause more problems in the body than value.*

5. *There is chlorophyll in any food touched by the sun.*

SURVIVAL LAW NO. 6

City dwellers and office workers who cannot get enough of the Natural sunshine directly should recognize that man can get the benefits of the sunshine through greens which are exposed to the sun. Chlorophyll-rich sprouts will pick up greenness when exposed to the sun even through glass. Both greens and the sun help control calcium in the body.

Points to Consider

1. Orange Juice is America's favorite breakfast juice. How in balance is that man who partakes of only this one kind of juice, neglecting the hundreds that exist in Nature? Nature is more than one big orange.

2. Chlorophyll, the bloodstream of the plant, is more a WHOLE in relation to the total plant than orange juice, the juice of the fruit, is. Can you see how man can benefit more chemically from chlorophyll than from drinking the same amount of orange juice?

3. Did you know that a bell pepper contains more vitamin C than the average orange?

CHAPTER ELEVEN

MAN THE ELECTRICAL BEING

Chlorophyll is possibly the missing link between earthly substance and human substance, and future experiments will make this relationship clearer. The essence of plant life, chlorophyll is an essence which man can use to build and rejuvenate his body.

Concentrated Sun Power

When we look at a plant under a microscope, the nature of chlorophyll becomes apparent. The sun and its heat have really made the green fluid dance. Here is the "concentrated sun power" E. Bircher described in his research. Here lies a vast energy resource captured in the leaf.

If we could "see" even closer, we would be watching a suspended cell life in constant movement. This movement is caused by the positive and negative influence of the cells working against each other. We would also be watching what science calls colloidal activity, "a natural activity of cell structure when basic chemical elements are broken down and become fit for human consumption."

A miracle of life takes place here in this colloidal electrical state. Because of the energy provided by the sun, a transformation occurs. Earth minerals, the "Dust of the Earth," are being made into food for man. These elements found in the plant life have been transformed and are entirely different than when they existed in rock form. The "Dust of the Earth," inorganic elements, are being REORGANIZED. Man, who can't live on the minerals as they exist in the rock, can live on the same minerals as the plant takes them from the rock and transforms them.

An Electrically Charged Body is a Healthy Body

These reorganized chemical elements work something like magnets in the human body. They repel and draw each other together at different speeds, and these different speeds of locomotion bring on different effects in various organs. It is the electrical state of energy in various organs which man uses to charge and recharge his body.

The day is coming when man will realize that he lives not only on the chemical elements themselves but the electrical nature of these transformed elements, this positive and negative force which makes natural, pure, and whole foods alive and growing.

When we cook wheat by steaming it, we can put it in the ground afterward and it will still grow. If you destroy the electrical activity which is going on in the cell structure, then you no longer have a live food. Any food that does not have electrical activity is not fit for human consumption. Few processed foods have such a charge.

The raw fooders get along so well because they are eating food which still contains this live force. One reason a person shouldn't have too much cooked food is because cooking breaks up the electrical activity found in the basic cell structure of plants.

Reasoning from a chemist's standpoint has developed so much today that they believe everything in the future will be explained in chemical terms. Yet again, the man who deals with finite electricity claims everything in the future will be explained from an electrical standpoint. I believe both these viewpoints will come together and science will see something in chlorophyll from both viewpoints.

Plants Provide Electrical Energy for Man

The day is coming when science will show that we absorb and use the plant's electrical energy as well as the chemicals themselves. We throw off chemical elements as much as we take them in. If we lived off them, why should we throw them away? Something in these chemical elements has been used, but what? The electrical charge makes them so valuable. A man dies and still has all the chemical elements he had a few moments before, but the electrical being is gone.

I stress that we live off the electrical and other components of the minerals rather than the mineral itself. The electrical charge in each mineral differs; we balance ourselves chemically on these charges. The chemicals themselves do not change. They have the same formula when we expel them as when we take them in.

Electrically Speaking

People should never eat the bottom of a vegetable unless they eat the top of it. Do not eat onions without eating onion greens. In many garlic products, greens such as parsley are included to neutralize the odor. Garlic is an underground vegetable and is balanced electrically by the parsley which grows above the ground. A different electrical influence exists in these two kinds of vegetables. The ones above the ground are positive and the ones below are negative; they have a different effect on the body entirely. Greens balance the negative electrical influence of underground vegetables because they have the positive influence from the sunshine.

We also find that the elements in the green tops of vegetables are refined in comparison to the minerals found below the ground. Those people who want to refine their bodies and live a more sensitive life can get the finer elements into their body much better by eating more of the tops of vegetables.

NEW AGE THINKING

In plants we recognize a positive and negative live force acting within each cell. Man has similar poles within each cell and within the WHOLENESS of his body. This electrical energy which is generated within the plant and comes to man through the foods he eats when they are pure, whole ·and, natural generates health in man. Randolph Stone, the father of polarity therapy, and other leaders such as Frances Nixon in the magnetic-electrical research field, are coming to terms with the vital nature of man the electrical being. Health's dependence on the energy currents which run through our body are one of the new age lessons of the new day.

SURVIVAL LAW NO. 7

Man, as part of Nature, must partake of electrically alive foods. Live forces exist in these foods which sustain and generate health. A natural energy flow exists within each cell of our body, and these electrical currents must be sustained. Few man-handled or processed foods have the electrical component left intact. Raw foods are live foods. I have seen the "dead" raised with warm goat's milk right from the young nanny.

Man — The Electrical Being

1. *The sun plays an integral part in the transformation of inorganic materials from the dust of the earth into materials fit for man's consumption.*
2. *Within each cell a positive and negative charge exists, and the constant movement within the cell life is attributed to the electrical influence of these forces.*
3. *It is the electrical state of energy in various organs which man uses to charge and recharge his body.*
4. *It is the electrical nature of these transformed chemical elements that makes our foods alive. Man was meant to eat live foods.*

• • • • • • • • • • •

"The Science of energy with its electronic and atomic research is the key to this new age."

                                                    Randolph Stone

CHAPTER TWELVE

MINERAL MAGIC

Iron, A Frisky Little Horse

Iron, one of the frisky horses in our body, is very rich in chlorophyll. Without iron, you're a deadpan; you have tired blood and you ARE tired.

Iron is called a frisky horse because it attracts oxygen and thereby brings oxygen to the body. You can breathe, and breathe, and breathe, but if you don't have iron in your blood from iron-rich foods, you will never get oxygen out of the air. When you are oxygenating the tissues well, toxic wastes which get in the way of the body's ability to function are also eliminated. With enough iron, the body is alert and a ball of fire, ready to accomplish whatever lies ahead. When your power and energy are gone, think of iron and its ability to attract oxygen.

It was Dr. V. G. Rocine, the professor who gave me much of my food knowledge, who always stressed that in order to have proper oxygen for our body, we had to have iron also.

Not All Iron is Fit for Man

When I first started working with liquid chlorophyll, I was considered a nut for talking about greens or salads. They threw tomatoes at me once. Now they have discovered tomatoes are good eating!

I have gone this "nutty" path for many years. We have finally gotten to the place where people are beginning to realize that there is something in foods from a nutritional standpoint, but they still must realize that we have a good iron (or any "good" element) for man's use as compared with a "bad" iron which he cannot use.

There is a difference between the good iron as found in chlorophyll and the iron in a ten-penny nail. If I were to grind up that nail in the finest form possible and give it to you in one hand, and then took the concentrated iron from chlorophyll and gave it to you in the other hand, which would you eat?

Or suppose I took an iron pipe and put it in one hand, in a nice, refined, clean, sanitary, BUT broken-down powdered form, and put iron-rich Idaho black cherries in the other hand, which would you choose?

Choose Clean Foods

Isn't there something that tells us to choose the clean material, the material that comes from plant life?

Yet although it is said in Genesis that we have the greens of the field for man's use, if you investigated the sources of many of our dollar products today, you might have a problem finding ingredients in a natural form that man could use healthfully. Iron in chlorophyll, however, is a natural God given iron, one that has been developed for mankind.

I heard a story recently about a doctor who recommends putting a nail in a glass of water, letting the nail rust, and then drinking the rust to get the iron. This is a lot of food foolishness!

Now, if he would have hammered all the nails in a tree so the fruit could get all the iron from the nails, and then he ate the fruit, he would be eating iron that was transformed chemically and electrically and composed properly for man's use. Again we see the "Dust of the Earth" has been transformed.

Henry Lindlahr, another nutritionist I admire, buried his tin cans at the foot of the tree so the tree could get the iron that it needed. This idea makes sense.

About Synthetic Vitamins

Many people today are eating foods which SKIP this transformation process. It is disappointing but true: Many products sold as foods are NOT foods. They come to us in forms we cannot assimilate. The consumer is in many cases eating minerals directly from the ground which have not been transformed for man's use. Some "smart" guy got the idea that chalk was high in calcium. So, we get chalk products as sources of calcium. But seeds and grains also have calcium in a high form. Which do you want, seeds and grains, or chalk? Who wants to eat the juice off a rusted nail?

Synthetic vitamins also skip this transformation step. The synthetic vitamin does not come through plant life, and I stress that vitamins must come through plant life, but for few exceptions such as Vitamic C. Foods cannot be polished, degerminated, or degenerated through heat or through addition of any materials (such as salt, brine, or spices). Man has never yet been able to take the inorganic from the earth and transform it as well as Nature does, BUT he keeps trying! You have to have foods natural! You have to have them real!

Our body has as much iron in it as found in that ten-penny nail. Imagine what we can do with chlorophyll when we constantly throw it into our body. We don't need a large amount of iron, but we have to have the amount we need.

Iron and the Thyroid Gland

We find that when iron attracts oxygen, it takes care of a lot of the work the thyroid gland should do. The thyroid is the oxygenating gland. We need oxygen to burn up the wastes in the body, to really cook our foods in the body so they are ready for assimilation by the blood.

More people have thyroid trouble today then ever before. People have broken down the thyroid by the way they think which destroys a good deal of our nerve force. With all the wars, political strife, and financial problems we have, we are not mentally at ease.

Luckily, the iron in liquid chlorophyll will rejuvenate these toxic thyroids by bringing more oxygen to the gland, but, of course, this would only be of value if a new path of thinking was also adopted as well.

Potassium Builds Muscles

Many greens are bitter because of the potassium in them. Potassium, one of the most alkaline of all elements in the body, is highly important. Consider how plants are built. The greater part of the chlorophyll is potassium. A great part of man is also potassium. Potassium makes up the muscle structure in the body; man is more muscle than anything else.

Bernarr McFadden, the champion cross-country walker, used to warn me, "Keep your muscles strong, because nothing in the body works well without the muscles working well." Without muscles, you're not going any place. You don't have a bowel movement without muscles or a heartbeat. You don't have the blood going anywhere without muscles.

There is nothing so swift and beautiful as a deer, nothing so swift and beautiful as any of the animals that live on the grasses and greens, and potassium gives the grace and beauty to these animals.

Some years ago they conducted an experiment at the university where they ran sodium salt through a heart and found it could only run fifteen minutes, but when they put potassium salt through the heart they found it could beat for three-and a half hours! The heart is a potassium organ.

A goal for better health is to keep the muscle structure clean and clear of any acids that may be stored there. 80% of our body is made of muscle structure. Can you see the volume of potassium that is needed for the proper functioning of this muscle system? Can you see why I recommend liquid chlorophyll as a heart support?

Magnesium

Magnesium, another essential element in chlorophyll, is found high in the tops of vegetables. Magnesium is wonderful in keeping ligaments and tendons strong yet relaxed. Here is one of the chiropractors' best friends. Magnesium helps build a strong straight back and aids the inter-vertebral structure as a whole.

Magnesium is stored here in this inter-vertebral structure and also in the bowel. If these tissues do not get enough of the magnesium, the tissues are unable to function.

Magnesium "hunger" symptoms are characterized by a toxic body, constipation, temper, burning and smoky urine, headaches in the lower forehead, hysteria, liver hardening, hepatitis, cramps in the legs, sleeplessness, fevers, and colic. This is just a sampling.

Dr. Rocine, the famed homeopath, tells us that magnesium is found heavily in seafood and unrefined foods such as rice and flour.

In controlled experiments on animals which lacked magnesium he reports kidney damage, loss of hair, and swelling (edema).

We should take advantage of the magnesium which is found in our green vegetables.

As with most chemical elements, our knowledge of them grows everyday. Perhaps the only safe way to deal with them, is to make sure we are getting and eating foods which are pure, whole, and natural. In this way we leave the balancing of them to Nature.

SURVIVAL LAW NO. 8

Man in the city today has difficulty getting the oxygen he needs from the air. A high iron intake, supplied from chlorophyll as well as our other survival foods, will build the red blood cell count and allow man to attract what oxygen there is in the air to his body.

The Magnesium Story Grows. . .

The Magnesium story seems to be growing by leaps and bounds, but we will likely never know all there is to know about any element.

Helene Kobzev, in the newspaper *Cosmos*, writes a revealing article containing much information about magnesium:

". . .Today. . .with the new science of biochemistry, scientists have come to realize how important is the long-known fact that our bodies are constantly generating tiny electrical impulses and discharging them. Long regarded as a curiosity of no great significance, these minute electrical charges have been learned to be an essential part of the processes of life and that without these electrical impulses, there could be no life whatsoever. And so today, we are compelled to recognize that if magnesium is the primary regulator of the electrical activity within our bodies, then magnesium is obviously of greater importance to health and life itself than anybody had guessed even 10 years ago.

"Recently," she continues, "much new knowledge has been gained about the role of magnesium in general metabolism. This mineral activates some 30 enzymes in the body; it takes an active role in the metabolism of protein, fat and carbohydrates; it influences the action of some of the vitamins and hormones. . ."

And so, the story goes on and on.

The Trace Mineral Copper

These three minerals, iron, potassium, and magnesium, are major elements in chlorophyll. There are many other minerals that are found naturally in chlorophyll too, and we could single out many to talk about which would stress the powerful nature of these ingredients. The story copper tells is especially interesting.

We need copper to build a good bloodstream and to keep the body from becoming anemic. Copper is a sunshine mineral, and this is why it is so high in chlorophyll.

One day, scientists will give copper a hard look from the biochemical standpoint and they will find out just how important it is for building the bloodstream. Scientists already tell us copper takes the longest of all the minerals to be made on the face of the earth.

To make this element requires the continual radiation of the sunshine.

Copper is one of our best conductors of electricity, and the most alive fluid we can put into our bodies.

There are some outstanding experiments involving copper. For instance, plant life can be kept in the greenest condition in the darkest room when reflecting any kind of light against copper on the plant! I have seen watermelons, casabas, and many other kinds of melons dried through the reflection of light on copper, and these dried foods had no germ life afterward.

A Dr. Fred Hirsch in Highland Springs resort in Beaumont, California, had a cereal on the market made of whole wheat and dates. No one else could produce a similar cereal because it would get so many bugs in it. By putting a reflection of light from a copper plate on it, he learned to avoid bugs in his packaged food.

I can't explain why copper reacts the way it does. Here is another secret left for the future. Here is another gift of Nature yet to be explored.

Summary of Chapter Five

IRON

1. *Iron attracts oxygen from the air.*

2. *Man cannot benefit nutritionally directly from the iron in the soil. He needs the plant go-between to transform this iron for man's use.*

3. *Many people today are eating foods which skip this transformation process. Synthetic vitamins and minerals are not substances that come through the soil. They act as drugs in the body. It is the synthetic Vitamin A, for example, that causes man problems with toxicity.*

4. *The iron in chlorophyll will help rejuvenate toxic thyroids by bringing oxygen to this gland.*

POTASSIUM

Potassium, the bitter element we taste in green leafy vegetables, is a muscle builder and heart support. A great deal of potassium is needed for the muscle system.

MAGNESIUM

Magnesium, another essential element in chlorophyll, is found high in the tops of vegetables. Magnesium is stored in the inter-vetebral structure and in the lining of the stomach. It acts as a muscle toner and relaxer.

COPPER

1. *Copper is just one of many finite trace minerals we find in Chlorophyll. Each of these trace minerals has a story to tell.*

2. *Copper builds the bloodstream. We find that vegetables can be kept in a green condition by reflecting light off a copper plate on to them.*

CHAPTER THIRTEEN

OVERALL DIET NEEDS

Man Needs Green Vegetables

As long as you live in this world, you will have some needs. Greens are one need you need most. You will be short many physical elements and greens are one basic adjunct to your diet making up for some of these shortages. Because of the foods we eat which are less than whole, pure, and natural, it's difficult to get all the trace minerals from the earth as they are found in ocean plant life, but chlorophyll will give us most of them.

Look at the foods in the marketplace. They have color, but do they have the elements? Do they have the colloidal activity? Are they electrically alive, or half-alive like some people?

A Healthy Living Regimen

After 45 years of nutritional counselling, I believe a good diet is six vegetables a day, two fruits a day, a good starch, and a good protein. Preferably, the starch and protein should be eaten at different meals. To make this healthy living regimen still better, a cup of liquid chlorophyll, made from the commercial concentrate, or a fresh green drink, can be taken at 10 a.m. or the first thing in the morning. For more suggestions concerning this regimen, see the appendix.

Four Food Groups?

There is much talk in general nutrition circles about the four food groups—meat and fish; milk; cereals and grains; and vegetables and fruits. The body does not need these groups in an equal proportion each day as much as it does need at least two large portions of different greens everyday. Never make a salad without Romaine lettuce, endive, or watercress, and don't forget sprouts as a clean source of greens! If you're talking about green juice, I don't mean taking a thimble full—you should have at least a half glass. It should be salivated carefully.

Keep in mind that 80% of our body is water and that we get this water best from fruits and vegetables. Having two greens each day, besides your other vegetables, should fulfill your body requirements.

Making Greens Appealing

Greens aren't very popular because they do not satisfy the need for sweets which has been so carefully cultivated in many societies, but because of their importance, find ways to mix them in your foods. I don't like greens at all, but if I ate just what I liked, I would be constantly eating Danish pastry and drinking Danish coffee. I would eat anything but a healthy diet! I have learned to eat not what I like, but what I have to eat.

While I do not like the greens, and especially salads, if you put a handful of steamed raisins in it, I will eat up that green salad with relish. Here is a nice idea for the children's salads—some cut up dates, or a little revived dried fruit mixed in, also helps to make a salad tasty. When preparing dried fruits, bring them to a boil, and turn off the heat until they cool.

It is the potassium in the greens that makes them so bitter, yet it is just such elements that make greens so valuable.

Eat it Raw

Another basic rule of health is to eat as many foods raw as you can. If you get 60% raw you'll be taken care of well. If you can eat a food raw, then consider eating it that way. In spinach, the oxalic acid is more difficult to take care of when the leaves are cooked.

I reluctantly and compromisingly tell you that you CAN cook! Certain foods should be cooked for your own good, foods such as potatoes, artichokes, and some cereals. However, there were no stoves in the Garden of Eden, and I am trying to give you a Garden of Eden without a can opener and without a French Chef.

Enzymes are in Raw Foods

We have to recognize that enzymes are one of our newest discoveries and a property which man doesn't know too much about. We think they lie within the protein molecule, that they have a lot to do with our life energy and have to do with our digestive energies. The enzymes are catalysts in the body which take care of all the other elements and foods in our body. Enzymes have the property of breaking foods down, of taking cooked foods and getting everything it possibly can out of them, and everything it can out of the raw foods.

Enzymes are found greater in the raw foods. We find that cooked foods destroy some of the natural enzymes. If this is the case, when we cook foods, we are also destroying some of the electrical energy—the live-food properties.

Some advanced thinking has never been revealed to man because we have never advanced ourselves in some of the universal studies that we should have advanced into.

Very shortly man will find that there is a radiation to all plant life and all foods that we eat, and food that has no radiation, and no electrical energy left in it, is no longer a live food. It should be called a dead food. This dead food lacks the enzymes for further activity when put in various parts of the body. I am convinced that live food is RAW FOOD. Raw food has the greatest enzymes—the greatest catalyst agents. I am positive there is a magnetic quality in enzymes. I am convinced there is an electrical quality in enzymes and a stored-up sunshine energy which is released in the human body.

Edwin Flatte, writing in the *Natural Nutrition Guide* makes some fine observations about enzymes:

"All fruits and vegetables in their natural state contain millions of living cells as well as vital enzymes. When food is subjected to intense heat for more than a few seconds, these enzymes and cells are devitalized.

"When the enzymes in fruits and vegetables are killed, our own body must be called upon to compensate by supplying the missing elements. This is an extraordinary drain on the body's reserves."

Further, "When reference is made to 'raw' food, actually it is not raw. The food has been cooked by the sun. When apples, pears, mangos, papayas, avocados, bananas, grapes, tomatoes, berries, melons, etc., have ripened by Nature, any further cooking or processing is unnecessary and harmful. Therefore, the term 'Live Food Diet' is a more appropriate description."

We find also that the enzymes are Nature's chemists. They bring fruits and vegetables to maturity. Under refrigeration, we can quiet their activity.

In an article by Kay Hallman in the *Health Gazette* entitled "Get Your Enzymes in Raw Food," the author reports that "Enzymes are sometimes referred to as 'ferments' because, as they die they cause food to become hot, to ferment, decay and spoil. That is why food processors cook, heat, pasteurize or add chemicals to food products - in order to kill the enzymes and preserve foods from spoilage.

"Each of your digestive glands and every hormone gland secretes a different and very specific enzyme, and each for a different purpose. At the same time, in order for the body to get the 800 different kinds of enzymes required for health, you must eat large quantities of raw enzymes from plants, whose molecules can be rearranged into hormones and digestive enzymes for your body's use."

Socking Greens to People

The raw fooder, however, comes along and thinks that all our foods aren't right and so we should go back to the raw foods entirely. There is no argument that it is the perfect way, but I find that most people can't take a lot of the natural foods in a RAW state because they have been away from them so long.

When you talk about "socking" greens to people, realize that you can also overdo it. If you started as a horse, ate greens all your life, your parents ate greens, and you stuck to greens, then you could live on grass!

However, anyone who goes to greens goes through such an extreme cleansing process that his body has to mold to this food and this takes years.

You could produce such heavy healing crises, elimination and transition processes, and changes in body tissue that you probably wouldn't realize some of these changes were more good for you than bad. Some people will lose weight and even turn yellow because of all the carotene that goes along with the chlorophyll foods.

Go slowly with greens. 10 to 15% greens, leading up to a maximum of maybe 20%. When you stop and think of all the other foods in our eating regimen, 10% is probably more than any one food in the whole daily routine. If you have your six vegetables a day, two should be greens—this is about 10%.

SURVIVAL LAW NO. 9

Nature cures, but she needs an opportunity, and opportunity means following the Laws of Nature. Nature cleanses and purifies the body very slowly. You must use good foods consistently over a period of time to expect some changes. Wonder drugs, although offering speedy relief, seldom cure, and those immediate reliefs are brothers to silent partners such as side effects and drug accumulations.

Following Nature's Laws

So far, we have been talking about the benefits of greens for the human body—for every human body. The Sunshine Liquid has a role in maintaining good health. When health is less than it should be, chlorophyll has proven itself as one of Nature's greatest healers.

Recently, a woman came in to see me really disappointed. "I ate a salad last week," she said, "and it didn't do me a bit of good!"

She had a lot to learn about the way Nature works. Nature cures, but she needs an opportunity, and OPPORTUNITY means following the Laws of Nature. Nature cleanses and purifies the body very slowly. The wheels of the universe grind slowly. You must use these good foods consistently, day in and day out, to get the good from them. The basic healthy eating regimen I have prescribed meets the requirements of Nature's Laws.

People fail to recognize the element of time in all natural healing. Wonder drugs, although offering speedy relief, seldom cure, and those immediate "benefits" are brothers to silent partners such as side effects and drug accumulations.

I have gone more into nature's laws in *Nature Has A Remedy*.

Vital Energy

When a person wants to get well, often in common practice the wrong measures are used. A person gets into an accident and is taken to the hospital. He is so weak that they just give him a little broth. The next day, he is a little stronger, so they give him broth and a few vegetables. In the evening, he gets a little stronger so they give him broth, vegetables, and a little meat.

The stronger he gets, the more they pile on and in greater combinations. The stronger he is, the more he must take care of. It would be wise, even when well and strong, to get on foods every once in a while which will not use up as much of your energy as you would have to use while eating extreme combinations of foods and many foods. Herein lies the value of a one day a week fast.

When we deal with getting well in a natural way, we consider VITAL ENERGY. It takes energy to digest potatoes. It takes energy to digest solid food, and hours of it. It takes little energy, however, to digest liquid chlorophyll. The bloodstream absorbs it in a matter of seconds.

When a person is fasting on water or on a juice diet in efforts to recuperate, he must consider that vital energy cures. Without vital energy we cannot eliminate. Without vital energy we cannot assimilate, digest, or do anything. So why lose your energy trying to digest foods when you are trying to conserve your strength?

"Simplify, Simplify, Simplify" — The Mono Diet

Closely related to the idea of regaining health through maintaining vital energy is the idea of the mono diet. We have heard of the "carrot juice lady" Mary C. Hogel, of Salt Lake City, who cured herself of a cancer by using carrot juice. Then we have Johanna Brandt who wrote the book on the grape cure. She cured herself of cancer on grapes. And then we have Herbert Shelton who believes he has cured people of cancers on diets of water alone.

Grapes, carrots, or water? I believe it is the mono diet, the idea that you need energy to get well, and foods like these provide a simple diet requiring little energy to digest and little digestive confusion.

The energy saved from digestion can be used for the repair, rebuilding, and rejuvenation of cell structure. These changes can only take place when we have the energy to do so. Most people have enervated their body to extremes through their work, mental efforts, and the eating of foods. The saving of this energy allows us to do this rejuvenation.

If I want to save a patient's energy, I send him to bed and give foods that take the least amount of energy to digest, yet foods that feed the chemical elements and infinitesimal trace minerals to the patient. The goal is to allow the energy that isn't being used to replace cell structure and build healthier cells.

If you can do this cell replacing at a low weight, then you can make the best body. At the end of a fast, you can do yourself the best amount of good because excess tissue also needs repair. For every five pounds of flesh, there are thirty miles of veins and arteries and a lot of fluids which have to be forced through these five pounds. It is said that a person 30 pounds underweight lives the longest life.

When you try to improve your health, I hope you will consider vital energy and the mono diet and the role chlorophyll can play in such a program.

Combating Drug Deposits

Drugs slow down the metabolism of the body and the recuperative and regenerative ability of the tissues. One report I see in the doctor's files says the average patient will spend 15 days in the hospital, but the average time will be 17 days for the person who has been taking drugs.

Two drugs that have caused trouble are penicillin and sulfa. They cause trouble because they are accumulative in nature and become part of the tissues. This keeps these tissues from becoming totally alive. The metabolism of these tissues is off and they cannot repair and rebuild as they should. These drugs also have many side effects and long range hidden time effects. I think the doctors will find that these drugs interfere with genes in the next generation.

People who have taken large amounts of sulfa have destroyed a certain amount of the tissues in the kidney.

The gastro-intestinal specialists are really concerned these days because antibiotics, namely penicillin and the sulfa drugs, destroy the friendly bacteria in the bowel. This friendly bacteria is called the acidophilus bacteria.

Greens Feed the Acidophilus Bacteria

If you're taking these drugs or any drug, feed the acidophilus back into the body. Above all, feed them with greens. One teaspoon of liquid chlorophyll in a glass of whey taken three times a day will help change the intestinal flora to a more natural one.

People are overloaded with drugs today.

Young people have gone through the psychedelic drugs such as marijuana, etc., and are looking for a way to clean themselves. Liquid chlorophyll, because of the distilling and refining process it undergoes, and because of the fine elements it contains, gets into the tissues, refines them, and makes them over.

OVERALL DIET NEEDS – SUMMARY

1. Greens are a basic adjunct to the diet. They make up for many of the shortages in elements we have everyday.

2. The best diet over the period of a day includes six vegetables, two fruits, one starch, and one protein. For more details about this Healthy Living Regimen, see Appendix D, Book I of this Magic Survival Kit.

3. At least two portions of green leafy vegetables should be in the diet everyday.

4. Because of the unpopular taste of many greens, creative and imaginative recipes to enhance their taste are called for. Some suggestions are provided in Appendix B.

5. Eat many foods raw when possible. Enzymes are found greatest in raw foods.

6. Enzymes are catalysts in the body. They care for the other elements in our body, breaking foods down and taking what they can nutritionally from raw foods and cooked foods.

7. When changing your diet to one where greens play a balanced role, expect changes in your health, some of them unpleasant at first. Gas and a loose bowel occur in many cases. A healing crisis where old ailments reappear and then are overcome also occurs after remaining on healthy foods for some time. During a healing crisis health reasserts itself with the help of Nature.

8. Ten to 15% greens, leading up to a maximum of 20% greens in the diet is sufficient for the average person. Two portions of green leafy vegetables a day is approximately 20% of the diet if following the balanced Healthy Living Regimen.

9. VITAL ENERGY is the net surplus amount of body energy Nature has to work with for natural healing of

the body. The more vital energy the body has, the better.

10. *A MONO DIET where we eat only one kind of food over a period of time causes less digestive confusion than eating many kinds of foods at once. The mono diet conserves body energy and is a program to follow from time to time when concerned about repair, rebuilding, and rejuvenation of cell structure.*

11. *Liquid chlorophyll, because of the distilling and refining process it undergoes and because of the fine elements it contains, gets into the tissues, refines them, and makes them over. Liquid chlorophyll washes drug deposits from the body.*

12. *Liquid chlorophyll also feeds the friendly bacteria in the bowel which are destroyed by many drugs in use today.*

CHAPTER FOURTEEN

CHLOROPHYLL ELIMINATION DIETS

Nature builds a good cell structure as well as it builds a bad one. To build a good cell structure requires good foods. While drugs do not build a good body, chlorophyll will.

Many times patients come to me ready to make changes. Some people say they are looking for a good doctor. I am looking for a good patient. These people have heard of the benefits of fasting and dieting and want a program where they can rid themselves of stored toxic materials. There are many programs we can recommend. All should be under a doctor's care, especially if you are considering going more than seven days on such a program.

Up until seven days, many people can take care of the diet themselves and in their own home, and we make suggestions here for such a home program.

Values in Chlorophyll Juice Diets

Why does chlorophyll deserve such a high place when it comes to recommending a juice diet? Here is a fluid that will help eliminate drug deposits. Here is a food that will help eliminate the effects of abnormal foods which have been building the tissues of your body wrongly for many years, the old toxic materials that have developed from hot dogs, dumplings, spaghetti. Here is a food that will help eliminate the effects from the sprays on our foods, and the after-effects of cold-tar products such as the artificial colorings and flavorings. These cold-tar products are prevalent in our canned, processed, packaged, and "enriched" foods. The Vitamin B which is used to enrich grains and breads is also a cold-tar product, as are materials in our cough syrups, alkalizers and aspirin.

DANIEL'S DIET — A Rigorous Cleansing Program

The most rigid elimination diet I can recommend is my DANIEL'S DIET for cleansing, building, and rejuvenating. This diet provides a complete rest for the stomach, intestinal tract, and bowel. It can be used one day a week or up to seven days a week without a doctor's supervision.

Use a teaspoon of concentrated liquid chlorophyll to a cup of water and take this every three hours. You should also take a bulk five times a day. A teaspoon of flaxseed, or the commercially-sold bulk-forming products. The bulk swells in the intestine and helps move things along.

For the FIRST THREE DAYS of the diet, take the juice every three hours and the bulk five times a day. On the FOURTH DAY, or when you start eating, drop the bulk to one time a day for a week. You can begin eating with meals of fruits and vegetables. The first meal should be a soft fruit such as peaches and apricots or sliced oranges. Shredded carrots. carrots slightly cooked, will also do for this first meal. Begin your regular health regime on the second day after you start eating.

Daniel's Diet will not stir up acids in the body but will eliminate them through the intestinal tract. On a water fast, toxic wastes are eliminated through the kidneys, bronchial tubes, and skin. You should have less odor on this program, a program which is actually both building and eliminating at the same time.

Suggestions for Less Extreme Diets

There are two other eliminating programs which I want to suggest, both of them a little less extreme. The first is suggested for four to seven days or two to three days as you prefer. We do not advise beyond seven days on your own. If you are elderly or if your energy is low, you can go on this diet for two days.

The basic juice on such a seven day program would be half-carrot and half-green, or one-third green. Parsley, rich in chlorophyll, and celery, because it is easy to take and an excellent source of sodium, are excellent ingredients. If these fresh juices are difficult to obtain, the liquid chlorophyll in the commercial form can also be substituted. This juice is made from alfalfa leaves. One teaspoon of the concentrate in a cup of water is equivalent to a cup of juice. You may add your carrot juice to this, or some other palatable juice.

For the FIRST TWO DAYS use a liquid chlorophyll drink made of carrots, parsley, and celery juice; a glass every three hours. Do not eat or drink anything else. The drink made with the concentrated liquid may also be substituted.

On the THIRD AND FOURTH DAY, add a little raw cow's milk or raw goat's milk and some solid food in the form of sprouts. These sprouts can be liquefied in the juice if you wish, or eaten separately. Have them without any kind of dressing while going through this particular diet. Sprouts are absolutely clean. They will not produce any toxic materials in the stomach and will promote elimination through the bowel. Sprouts will not spoil in the bowel like cooked food or meat will.

When Eating Meat . . .

Note that cooked foods, including vegetables, spoil in the bowel quicker than raw vegetables. Meat is the most putrefactive and spoils quicker than any food you can put in the bowel. Anybody who eats a lot of meat should eat a lot of greens. The greens will help keep these foods from putrefacting and spoiling. A goal of healthy living is to get such foods out of the body before they putrefy in the bowel and spread these toxic poisons through the system.

On the FIFTH AND SIXTH DAY add some of the more solid

vegetables like the squashes to your diet. Banana squash and zucchini are excellent bowel cleansers while yellow neck, summer squash, and pumpkins are also good. Use them either cooked or raw. Also add shredded beets for the liver and some watercress which is so high in potassium.

On the SEVENTH DAY begin eating salads made with Romaine lettuce, Red Tinged or Bibb, but with no dressings.

On the EIGHTH DAY begin your regular diet as we have enclosed in the appendix. This diet has enough roughage for you to go right into your regular diet. The first day after the diet, however, use no starches. The second day after, you may add them.

The drinks should be taken every three hours and on the last three days, the salad three times a day. When you begin the salads, continue taking the drinks. In exchange for a salad for breakfast, substitute one solid fruit. You can also have sprouts on these last three days for breakfast if you so desire.

This is still a chlorophyll diet, harsh in its cleansing effect. It is a little more extreme than a diet just of vegetable juice.

In a case where you do not wish to go the full seven days, stop the elimination by adding some of the nut butters and avocado dressings. Put avocado in with some of the drinks to make them more like a milk shake, or add a little honey.

For a more palatable juice, add a little apple or pineapple juice along with the carrot and green juice on the fifth, sixth, and seventh day. These juices should be fresh. They should be salivated carefully in the mouth at a slow pace.

The Once-a-Week Fast Program

If I had my way and could stay in bed one day a week, read a book, watch a little TV, and everyone left me alone (and my dog never said a word), I would take a fast on water or a diet on liquid chlorophyll one day a week. It is unfair to the body to take such a fast or diet and still put out energy. This day must be a complete day of rest—mentally, physically, and physiologically (foodwise).

Exercise

While you are on this diet, skin brush,* use a slant board, and do some exercise. Go to bed, get up and walk, and then go to bed again. Two hours down, and one-half to one hour up. If you take any sunbaths, do not overdo it. When you start sunbathing, begin with no more than 15 minutes a day.

The Role of Enemas

During the diets, it is possible for many people to have terrific bowel accumulations which will be necessary to eliminate. An enema may help. I recommend using a liquid chlorophyll enema (see "Bleeding Bowel" section for details). This enema will help bring the bile down through the liver and gall bladder. You can take such an enema once or twice a day if you are uncomfortable with gas.

Skipping Lunch

A modified diet which is less extreme still and yet especially good for elimination would be to only eat meals in the morning and evening and take liquid chlorophyll drinks in between for cleansing the body.

Sprouts for Elimination Diets

One of the nicest chlorophyll diets you can have is built around the sprouts. Sprouts act as solid food in an elimination diet, yet do not interfere with the on-going cleansing process. You can enhance the sprouts with an avocado sauce, also high in chlorophyll—green runs through it. To make sprouts palatable, have them with nut butter dressings, but keep in mind that nut butters will put weight

* NOTE: Skin brushing entails brushing the skin with a dry vegetable brush. This is one of the finest baths you can have. Use a skin brush before water bathing each day.

on. Other ideas for enhancing sprouts are using a tomato sauce dressing thickened with Agar-Agar or arrowroot, or in a gelatin mold with mint added. Mint is also high in chlorophyll, as are many other herbs. These suggestions are for the average person going through a health-building chlorophyll elimination diet.

More information on cleansing is in my DOCTOR-PATIENT HAND-BOOK dealing with THE REVERSAL PROCESS and THE HEALING CRISES through elimination diets and detoxification.

CHLOROPHYLL CLEANSING

1. *Chlorophyll deserves a high place in eliminative diet programs because it is a fluid which helps clean the cell structure of the body. It has vital minerals to help build these structures with new cell life.*

2. *Daniel's Diet and various other elimination programs enclosed in Chapter Seven can be used without the benefit of a doctor if these programs are set up for SEVEN DAYS or less.*

3. *Exercise, rest, and enemas should play a part in any eliminative program.*

4. *Sprouts are an excellent food for eliminative diets. They provide a soft bulk for the intestines yet do not interfere with the cleansing process.*

CHAPTER FIFTEEN

HOW TO GET CHLOROPHYLL FROM FOODS

Many years ago I spent time in Bircher-Benner's sanitarium in Zurich, Switzerland, where I watched them experimenting with early kinds of juice machines. Even before this visit, however, I was involved in drawing the chlorophyll from plants by one of the simplest methods.

Chopping Leaves

I would take the leaves, chop them up diagonally across the arteries, and put these finely chopped leaves in distilled water. After an hour, the water would be green from the bleeding arteries. After the greens have been in the water for an hour or two, squeeze them through four or five layers of cheesecloth, and then drink the water. If you have no juicer or electricity, this is one of the finest ways to get your chlorophyll—cheap, easy, and definite.

Vegetable Juicer/Liquefier

What I have described so far is the most natural and simplest method for obtaining liquid chlorophyll. The next step up is using a vegetable juicer or liquefier. With the liquefier, separate the pulp from the juice through cheesecloth or a strainer.

Concentrated Liquid Chlorophyll

The next step up is for those who don't have a juicer/liquefier. They can get the liquid chlorophyll commercially in a concentrate. It also comes in capsule or powder form, and in tablets.

In this book, the remedies for the most part are geared to the person who will not have the time to make the fresh juices for himself, and for the working person or teacher who would like to take the concentrate to work. The concentrate is especially convenient because it will not spoil as most fresh juices will. It can be diluted, one teaspoon to a cup of water, as you need it. Many brands exist but I prefer a magnesium chlorophyll. I believe the magnesium chlorophyll is the most natural and therefore best-suited for the human body.

Juice Lifespan

Regardless of the type of fresh juice which is made, drink it immediately. Never keep a vegetable juice more than six hours. If you put a nail out in oxidizing weather, it becomes brown and rusted. Oxidation takes place in vegetable juices because they are also high in iron—although, of course, of a different kind than the iron in that nail.

These juices should always be ingested very slowly—never gulp them down.

N. W. Walker, in his book *Raw Vegetable Juices*, states that, "Vegetable juices are the builders and regenerators of the body. They contain all the amino acids, minerals, salts, enzymes, and vitamins needed by the human body, provided that they are used fresh, raw, and without preservatives, and that they have been properly extracted from the vegetables."

It's important that these vegetables be grown on healthy soil without artificial fertilizers and sprays.

SURVIVAL LAW NO. 10

Fruits and vegetables today are much bigger than they were in their original form. They have been watered down greatly. When they were small, to take in a concentrated juice was unnecessary. Because of the changes, it is necessary to include ample fruit and vegetable juices in the diet to get needed minerals and trace minerals.

NOTE: Ways to introduce additional chlorophyll foods into the diet are listed throughout the appendices.

CHAPTER SIXTEEN

A VAST ARRAY OF SOURCES

There are many vegetables and herbs in God's Garden and we are likely only scratching the surface when we discuss the properties and benefits of chlorophyll. Cabbage and green kale have been used successfully as packs, while miracle properties have been attributed to comfrey, a dark green leaf so highly praised in H. E. Kirschner's *Nature's Healing Grasses*.

I have reports of a local Escondido, California dentist who says he has improved his ulcers of the stomach with comfrey. Adding comfrey juice to flaxseed tea soothes the stomach.

In any case, I hope the remedies described here and the exciting results you can receive will start your own chlorophyll search and inspire you toward a greater understanding of the value of chlorophyll.

Aloe Vera Suggestions

A source of many new remedies will be the cactus family. Because it is a sunshine plant, it has a lot of chlorophyll as well as many unknown healing agents. I am positive cactus remedies are wonderful for diabetes.

Aloe Vera—a member of the Lily family and a close relative—is the only plant that can take care of x-ray burns on the human body. If they have found this so, why not investigate Aloe Vera further?

Some exciting information is already coming to the surface about Aloe Vera. For skin burns and rashes, apply in a lanolin salve. This is made by mixing equal parts of lanolin and Aloe Vera gel (as sold in your health food store). For relief of muscle cramps, apply the gel or put on the plant's mucilaginous liquid directly. In packs for arthritis, run the pulp of the plant through a blender and then apply on the skin. Cover with a damp cloth and then a dry one. It has a wonderful effect for relieving arthritic pain.

It is also a wonderful laxative. Remove the skin and drink slowly three or four half-inch squares well-blended in any kind of juice.

Aloe Vera is also outstanding for aiding colitis. Mix ½ teaspoon of

Aloe Vera

Aloe Vera gel with one cup of flaxseed tea made by the method described under "Bleeding Bowel." Retain for 10-15 minues before expelling.

For ulcers of the stomach, add a teaspoon of Aloe Vera gel to flaxseed tea. It is best to use the gel rather than the liquefied plant itself because the pulp might irritate the ulcer.

Aloe Vera also makes an excellent after-shave lotion. Apply the gel or the mucilaginous liquid as found in the untreated plant.

I tell you this information to whet your appetite for some of the finer forces existing in Nature. We can discover them when we get close to her. Perhaps the future of man's inventions lies in discovery of the world around him which is natural, pure and whole, rather than the world which is artificial and plastic.

Wheatgrass

An article in the *National Health Gazette* reports that wheat is one of the best grains to store for survival because it is so high in nutritional value. Wheat picks up 92 of the 102 minerals in the soil. It contains calcium, phosphorus, magnesium, sodium and potassium. Wheatgrass, the article continues, is a complete food. It is high in protein, and a rich source of essential enzymes which the pancreas needs to aid digestion of starches. Because of its chlorophyll content, wheatgrass is one of the finest blood builders and body rejuvenators I know.

How to Grow Wheatgrass

To grow wheatgrass, plant whole wheat which has been soaked for a number of hours in a tray or box filled with dirt one inch deep. Sprinkle the whole wheat grains on top and then cover them with about one-quarter inch of soil. Dampen the soil each day just enough so the grass will grow. When the grass is about five or six inches high, chop it off and make the juice out of it.

You can put the wheatgrass through a hand grinder and separate the juice from the pulp. You can also run it through a juicer or use your liquefier and then strain the juice through a few layers of cheesecloth. When using the liquefier, add equal amounts of water.

Bill Estes, a physical therapist of Hidden Valley, California, suggests that this pulp makes an excellent poultice for drawing out the toxic materials in muscles and promoting the healing of wounds. Place the pulp on the skin in a pile and then cover with a moist cloth. Cover this with a dry cloth. You can leave this on overnight. This wheatgrass pulp is also reputed by Estes to bring relief to arthritic joints.

Wheatgrass is a food you can have year round. You can plant it and keep it in the pantry or under your bed. Bring it out into the sun before harvesting.

I don't believe any one food is everything, but wheatgrass is a fine adjunct to any diet and a drink for those in special need. You can have a glass up to three times a day. Make sure you salivate it

Wheatgrass can be grown in many ways.

carefully in the mouth before swallowing. Drink it very slowly.

I have used wheatgrass for more than 30 years in cases of anemia, muscle debility, and as a general health builder.

Replenishing the Soil

In an article in the *When* newspaper, Dr. Ann Wigmore reports that soil which has been chemically treated can be returned to a natural state by the following procedure:

"Wet your soil thoroughly. Spread soaked wheat seed thickly on the surface. Cover this seed with large sheets of plastic. Place stones on edges to hold the plastic in place. Do not uncover until growth underneath plastic is two or three inches tall. Keep the earth thoroughly moistened until the shoots grow about seven inches tall. Turn this growth under. Wait two weeks for this turned-under stubble to rot. When it has disintegrated, break up the soil and plant again. Three or four of these plantings will make your garden organic.

Algae — Another Food for the Future

Sources of chlorophyll are yet undiscovered by many people, and our concept of good food will likely change in the coming years. In Japan, where they have a problem getting and growing foods, they grow "algae." Algae is a basic food. Plant life starts out in ponds, and germ life gets its start here. Yet Algae may be one of the foods of the future for man too.

Because they have so many people in Japan and such a need, they have learned how to grow algae on roofs and keep it sanitary. Usually algae grows in stagnant pools; it is the green scum which grows on top. The algae can be kept sanitary if the water can be kept clean.

I have seen algae used for many purposes. They make it into ice cream and bread which is as green as grass when baked.

In Palm Springs one nutritional outfit recently gained approval of the Food and Drug Administration to market a product which is basically an algae. It has many trace minerals in infinitesimal amounts which have been nurtured by the action and reaction of the sun and water.

The discovery of other new foods and new uses of existing foods will be a boon to man's health and will be coming in the future.

SURVIVAL LAW NO. 11

The new age man will know the weeds and grasses of the field. These original foods come to us undiluted and untouched by man. The following list is just a sampling of wild weeds, herbs, and grasses available as food for man:

Angelica	Clary (sage)	Mustard
Anise	Comfrey	Nasturtium
Balm	Chicory	Oregano
Basil	Tansy	ParsleyRue
Bay	Dandelion	Rosemary
Borage	Dill	Sage
Burdock	Fennel	Savory
Burnet	Geranium	Winter Savory
Caraway	Horseradish	Sorrel
Catnip	Juniper	Tarragon
Chervil	Lavender	Thyme
Chives	Lovage	Caraway Thyme
Cicely	Sweet Marjoram	Watercress
	Mint	Woodruff

The foregoing list of wild weeds and herbs was taken from Lesson 20 and 33 of my *Home Study Course* which covers the use of herbs and herb teas. Outstanding works on herbs and the grasses abound. Some are *Eat The Weeds* by Ben Charles Harris, *The Healing Power of Herbs* by May Bethel, and *Nature's Healing Grasses* by H. E. Kirchner.

The beginning survival student should browse through the books on health subjects which are offered by your local health food store.

Malva — The Case for the Wild Weed

One of the best greens for controlling calcium and also discharges from the eyes is Malva, a weed which grows wild throughout the United States and elsewhere.

When I was in the Hunza valley, I found that many cases of eye discharges could be improved using Malva as a vegetable juice remedy.

I have a report which states that there are 50,000 units of Vitamin A to a pound of Malva. Because it is such a green leaf, this weed is doubly powerful as a healing agent.

Malva is tasty and can be used in salads as well as juice, but because we see so much of it in the Spring, we ignore it. We buy spinach and pay so much a pound for it, so it must be good food! You wouldn't eat a weed anyway, although a weed like Malva might be more valuable than much of the food you can buy in the store!

Cleaning Vegetables

Malva and some greens have to be washed well because there is buglife and bacteria on them. To clean vegetables, mix one teaspoon of Purex or Chlorox in one gallon of water and let the vegetables soak for five minutes. Take the vegetables out and rinse them off in plain water.

There is also colloidal sulphur which you can buy in a health food store. This will help get rid of this worm and germ life, and is excellent for washes.

Neglecting Greens Costs Money

There are also many other places where we can get chlorophyll, places most people overlook. Get the green from pea pods! Throwing away the pod is a great waste. When you buy sweet peas, you should be able to eat the pods also. Consider making a broth out of them or vegetable juice. I have tasted pea pod juice which was some of the sweetest juice you would ever want to taste.

Nasturtium leaves (a flower) are also an excellent source.

One of the greatest crimes is throwing away the inner leaves of the cauliflower which are one of our highest foods in iron and potassium. We take only the bleached inside plant life which has been held away from the sunshine (which gives us the calcium control). The cauliflower greens are excellent in juice or as a cooked vegetable.

I am inclined to believe that lima bean pods would also be of value.

Sprouts

Another way of getting chlorophyll into the diet is through sprouts, one of the survival foods you should learn more about. When you dampen seeds they begin to sprout. These sprouts we consider one of the finest of all foods.

The power in the tiny seed is overwhelming. Parrots and turkeys can live solely on sunflower seeds and they will remain in perfect health for years. There isn't any one food you can use so well for birds.

Alfalfa sprouts should be soaked overnight.

When grown into sprouts, these sunflower seeds contain many natural elements, including the potassium and iron we have said so much about.

Sprouting Sunflower Seeds

To sprout, sunflower seeds should be planted in a tray or box in damp soil. Pepper the soil with sunflower seeds in the shell one half inch into the soil. Cover them over with damp soil and keep it damp. In about three days—sprouts! When they are about 3" high, place them outside (or grow them outside) so they will turn a dark green. You do not eat the mother seed the sprout has grown from. Cut the seed off.

Sprouting Alfalfa

Alfalfa sprouts can be grown in a quart jar. Cover the opening with a few layers of cheesecloth. Soak the seeds for a number of hours and drain off the liquid which you can drink. Lay the jar on its side and rinse with water two or three times a day, always draining off the water. They should be grown in a dark place and taken to the sunshine after they are grown. One large tablespoon of seeds fills such a jar with sprouts.

Minerals Have Colors

Someday in the future, man will test seeds in the laboratory before a spectroscope and find that seeds have the most complete reading of minerals found in the earth, including the trace minerals.

When seeds are tested on a spectroscope from a color standpoint, we realize that all the beautiful colors we find in the parrot—the blues, greens, and the yellow and red feathers—are all fed by the various elements in the sunflower seeds the bird eats.

Each chemical element has a color and they produce color in the parrot. We feed the various hues, shades, pigments, and organ colors in the bird with the colors which are found in these seeds.

Man also has color. We have different colors in the various organs

Two day old alfalfa sprouts.

Five day old alfalfa sprouts.

of the body. The gall is yellow, the bile is yellow, the stomach is white, the heart is red and a fluid in the eye is black. The colors are made from the elements and these colors have to be continually fed by the minerals in our foods in order to keep our body bright and healthy.

The value of sprouts is emphasized by events in the bird cages at the San Diego Zoo. A while back, they were spending $15,000 a year for lettuce to get fresh greens for the animals, but they found that many of the offspring were not in good health and did not live. Even though this lettuce was proving ineffective for solving the zoo's problem, the keepers still felt the live greens could help remedy this health problem.

Since they began using sprouts instead of lettuce, the zoo now finds that the animals have few problems with procreation or raising healthy children. The animals are healthier now than the zoo ever had before!

Ducks and geese from Canada usually seek lakes throughout America to rest on their way South, but since the introduction of the program of offering fresh sprouted greens at the zoo, the keepers find that many of these wild birds skip the lakes and come to the zoo park to partake of these live foods.

Now how do these birds know the value of these sprouts and why should they be attracted to them? This is something to think about.

Up in Fresno they are using oat sprouts instead of their regular feed for horses and getting wonderful changes in the health of these animals.

Alfalfa

Alfalfa is just about the most alkaline plant that we have. Many diseases exist in a toxic-acid overloaded body and alkalinity helps balance these conditions. Linda Clark writes that alfalfa is one of the most complete and nutritionally rich of all foods tested. In addition to a fabulously high potency of vitamins as well as minerals, it is high in protein and contains every essential amino acid. Richard Lucas, in *Nature's Medicines*, writes that the protein content is 18.9 percent as compared to 16.5 percent in beef, 3.3 percent in milk, and 13.1 percent in eggs. It also reports eight known enzymes in it.

Alfalfa has a deep root structure which if stretched in a straight line would measure from 125 to 250 feet, various reports indicate. In having such an extensive structure, these roots pick up many of the trace minerals that build the bloodstream and organs. Alfalfa has ten times the mineral contents of most grains.

Its phosphorus, chlorine, silicon, aluminum, calcium, magnesium, sulphur, sodium and potassium are in just the right proportion to build strong bones, sound teeth, steady nerves, firm muscles, a regular heart beat, healthy organs, a fine posture, and keen mind.

The homeopath, who produced in a healthy person symptoms of disease in order to build up a resistance to disease, used lucerne as one of his main remedies. Lucerne is alfalfa.

Alfalfa sprouts make a soft bulk for the intestinal tract. It's not going to make this system overwork. These sprouts, we find, are also rich in the vitamins A, B, C, E, U, and K. H. E. Kirschner, in *Nature's Healing Grasses*, says these sprouts contain vital cell-building amino acids. He reports they contain 150% more protein than such grains as wheat and corn.

A meal made by grinding alfalfa seeds is used as a basis for many high potency vitamin tablets. A tea—very beneficial for arthiritis and rheumatic sufferers, is made by steeping one tablespoon of seeds in a quart of boiling water. Take this three times a day. Alfalfa leaves are a popular source for the commercial liquid chlorophyll concentrates.

Herbs

Herbs are the most potent of plants. The majority are the same now as they were thousands of years ago. Our familiar vegetables of today started off as herbs, but they have been conditioned by man so much that they have lost in strength what they have gained in size and mildness.

Herbal leaves, barks, roots, and seeds have vital essences of importance to our health and resistance to disease. They are dissolvers, openers of obstructions, eliminators of dead elements and waste materials. They cleanse and purify the whole body. Any inflamation, infection, or putrefication is resisted by them. They are an antidote to poison; a natural relief from aches and pains.

Genuine herbal methods normalize body chemistry and control cell metabolism. They are a tonic to the whole system.

Herb Teas

Herb teas are excellent food adjuncts, having many of the herbs' healing qualities. I have never found a person who did not need silicon, the magnetic element. Did you know there is a tea made from the stalk of a cereal grass that is highest in this essential element - oatstraw? The tannic acid in ordinary tea is comparable to a drug which may be more injurious to the body than coffee.

Become acquainted with the different herb teas and use one or two a day, combining any two according to your liking. Sweeten them with juice concentrates, such as apple or cherry, or natural fruit syrups. Use honey, too, but sparingly. Honey is a concentrated sweet and should never be used in excess of three teaspoons a day.

More herb information is in Appendix B of this book.

CHAPTER SEVENTEEN

CHLOROPHYLL REMEDIES

The Secret of Getting Well

The secret of getting well, if you have lost your health, is to eat foods that will build your blood, get the blood where it is needed through exercise, and rest. In this book I discuss chlorophyll which is one food which will help build your blood, and add to it the chemical elements it needs.

In the following pages, you will find many successful treatments which have been aided by chlorophyll. I want to stress again, however, that these are not treatments which should be applied without the advice of a doctor for the most part. Further, recognize that the potential of chlorophyll is unleashed to those people who are following Nature's ways and leading a life where they are trying to live within the finer forces of Nature.

The benefits of chlorophyll will be slim for those who continue to live on coffee and doughnuts or many of the ghost foods that line our shelves today. If you are living on a diet which is breaking you down, don't expect liquid chlorophyll to build you up. You must use every force necessary in your diet and daily living for healing. The secret is learning what a healthy way of living is. At that point, Nature's gifts will be yours, including the potent force which is in chlorophyll.

Remedial Effects of Chlorophyll

Builds a High Blood Count
Provides Iron to Organs
Counteracts Toxins Eaten
Improves Anemic Conditions
Cleans and deodorizes Bowel Tissues
Helps Purify the Liver
Aids Hepatitis Improvement
Feeds Heart Tissues Iron
Regulates Menstruation
Aids Hemophilia condition
Improves Blood Sugar problems (diabetes)
Aid in Asthma Improvement
Increases Iron Content in Milk
Improves Milk Production
Helps Sores Heal Faster
Eliminates Body Odors
Resists Bacteria in Wounds
Cleans Tooth & Gum Structure in Pyorrhea
Improves Nasal Drainage
Slows Nasal Drip
Lessens need for Underarm Deodorizers
Eliminates Bad Breath
Relieves Sore Throat
Makes Excellent Tooth Surgery Gargle
Benefits Inflamed Tonsils
Soothes Ulcer Tissues
Soothes Painful Hemorrhoids and Piles
Aids Catarrhal Discharges
Revitalizes Vascular System in the Legs
Improves Varicose Veins
Reduces Pain Caused by Inflammation

When I say, "get next to the finer forces in Nature," I am asking you to open your eyes and really see. This is quite a request when so many around are doing otherwise.

Once when studying how to get the greatest amount of good from liquid chlorophyll, it was learned that four in the morning would be the best time to harvest. They found that in the daytime, much of the liquid chlorophyll was down in the roots but was in the leaves at night.

One day, we're going to harvest our foods during the night or early morning, and that will be paying attention to the finer forces of Nature! You may be surprised to learn that plants do most of their growing during the night.

In the following remedies which I give for chlorophyll, I would hate to think that someone who had a serious condition that a doctor should be taking care of was trying to apply some of the simple remedies which we are giving here. You should have a doctor that takes care of you at all times. In the following pages you will find tested and proven remedies which your doctor may be interested in. However, I am not saying these remedies should take the place of more serious treatments for serious cases which demand a doctor's attention.

Experiments with Chlorophyll

An experiment using rats was conducted. Two groups were used, one group having chlorophyll in their diet. These animals were put in water to swim for their life, and the group that did not have chlorophyll could only swim for one hour, while the other group survived for three hours. In another experiment with rats, the blood was extracted, and liquid chlorophyll put in its place. The rat went right on living.

It was Emile Burgi of Berne, Switzerland, who did a lot of work with chlorophyll and said it was excellent for influencing anemia of various kinds, for improving the action of the heart, and for reducing blood pressure. He claimed it stimulated peristalsis, improved the intestinal action, and was a mild diuretic (it makes the kidneys eliminate).

The Secret of My Work — A High Blood Count

The secret of my work is building a high blood count which means having the extra red blood cells to deliver the chemical materials to the various organs of the body so they can repair and rejuvenate themselves.

Sick people very seldom have a good blood count. For this reason, doctors often have the count taken as one of the first tests they conduct as an indicator that something abnormal is occurring. For good health, build a blood count that is not only normal, say 4,500,000 but one above this.

Altitude Builds a Blood Count

Mountain dwellers who live at an altitude of 10,000 to 12,000 feet have a blood count of 7,500,000. These men teach us that the old adage, "Go to the hills for thy strength," has much meaning.

Mr. Gassanov, the 153-year old man whom I visited in Russia during 1971, had a 6,500,000 blood count according to his personal physician. It takes a blood count of this kind to live a long, healthy life.

A Higher Blood Count Means More Nutrients

A higher blood count means the privilege of taking more of the chemical elements from your foods. Even if the foods are anemic and have only a few trace minerals and chemical elements, with a high blood count it is possible to get more nutrients from these foods.

Many people have very poor circulation and don't get the blood around the body as fast as they should. If the blood is anemic and not moved around, then you are deadly anemic in the tissues and various organs.

A high blood count and a high iron content in the blood will mean that even if the circulation is poor, you are going to have a better tissue and better repairability.

Greens Go First

The importance of greens is universally known. In Switzerland, they say you should start the meals with them. They believe this builds up the white cell blood count and takes care of all the toxic materials that may be eaten at the meal. They find you have a much better blood count after you eat when you start the meal with a salad.

People who have sedentary jobs and are not exercising must work on having as high a blood count as possible because their circulation will be poor. Because greens build a high blood count and bring iron to the blood, every person should make them an important part of the diet.

Raising the Blood Count in Anemia

There is nothing that stimulates the formation of red blood cells better than chlorophyll-rich drinks.

One of our most unusual cases showed me this value. Mr. Pete Maloff, one of the leaders of the Doukhobor colony in Canada, brought his daughter to us. Mr. Maloff was a third generation vegetarian, and because his daughter was a fourth generation vegetarian, this way of life was so instilled in her that she said she would rather die than eat meat.

She was ill with pernicious anemia. The Mayo Clinic advised they would have to give her shots of liver and iron and beef products to build up the blood. She refused this treatment.

When I find people so committed, I must help them at their state of consciousness.

I started giving her an excessive amount of liquid chlorophyll over a period of three months, a teaspoon of the concentrate eight times a day. We put one million red blood cells in this little lady's blood count in just one month. She went from 2,800,000 to 3,800,000 by just using liquid chlorophyll. When she left her blood count was 4,500,000 which was almost up to normal.

The girl regained her health and became very active and vigorous. She married and has beautiful children. She outgrew the problems of her early life, a time when she had difficulty keeping up with others.

Iron is one of the elements we need to overcome anemia. Anyone who is anemic is tired, fatiqued, can't move. They have no energy and their concentration is gone.

There are many times that iron and blood shots do not overcome an anemic condition as well as liquid chlorophyll. Nothing I can tell you will build blood as fast. It is one of the best blood builders we have.

As Internal Bowel Cleanser

The bowel is the first line of defense against any illness or poor state of health. Small weak spots in the bowel wall form pockets called diverticulae. These pockets fill with toxic materials which interfere with the health of every organ because they interfere with the metabolic rate.

To improve the metabolic rate of digestion and of the various organs and glands, make sure the bowel is kept clean. The bloodstream is no cleaner than the bowel, and the metabolic rate is in direct proportion to the toxic settlements there.

The metabolic rate of any tissue in the body is the movement at which every organ is functioning. When these organs function at a low speed or throw out an under-amount of hormone or digestive juices, then the metabolic rate of that organ is too low. (Sometimes too much hormones can be thrown out causing a high metabolic rate, but this is not usually the case). Tissues can be overactive when irritated with extreme acids, while toxic materials in the stomach can make many organs underactive (which is the same as having a slow metabolic rate).

Cleansing with Green Tablets

When we use chlorophyll we help feed the bowel tissues and deodorize them. When we use alfalfa and chlorophyll tablets we are providing the bowel wall with a fibrous material which this muscle wall can work against the toxic settlements in the diverticulae. These fibrous materials get into the pockets and help clean them out while the chlorophyll in the tablets deodorizes at the same time. The fibrous material also helps to develop the tone in the bowel wall. This treatment is especially good for constipation.

The average person can take four or five alfalfa tablets with each meal, and people with extreme toxic conditions can take up to ten tablets and only cause good to happen. Make sure you crack them once before swallowing and use along with water or juice or blend in a liquefier. Other values of alfalfa are explained under the headings "Sprouts" and "Alfalfa."

Many doctors believe diverticulae are incurable. I consider them an inherent weakness which may be incurable, but we can keep the low-grade infections in these pockets clean by daily ingestion of these tablets.

Preventing Bowel Accumulations

To further improve conditions in the bowel, a good digestant such as hydrochloric acid tablets can be taken, and also a pancreatic substance as a digestant. These help digestion and prevent toxic buildup in the tissues of the bowel.

The pancreas is a gland which digests starches and sugars. Most people don't have a strong pancreas and so undigested starches many times accumulate in the body and cause toxic settlements.

Hydrochloric acid tablets are not easy to prescribe, but anyone can take one with each meal. In cases of hyperacidity of the stomach, do not take any. However, even if you have a stomach that has a metabolic rate that is overactive, you can still use pancreatic materials in all cases.

When we take care of the bowel, we have less problems throughout the body.

Our Body's Great Detoxifyer, the Liver

Just as the stomach is considered a sodium organ, the liver is in need of more iron than any other element. One-fourth of all blood is in the liver right now, and the organ needs much iron to purify the blood which circulates through it. It is in the liver where much of the

Jensen - chlorophyll

94 - wheatgrass pulp - wound healing,
      arthritic joints

95 - replenish soil - wheatgrass

99 - sunflower sprouts, alfalfa

102 - alfalfa seed tea - arthritis

108 - iron - blood - chlorophyll

109 - 110 - constipation - alfalfa and
      chlorophyll tablets

111 - detoxify liver - chlorophyll (iron)
      dandelion tea

toxic materials are removed from the bloodstream. This is why the liver is called the body's great detoxifyer.

A toxic liver demands large amounts of liquid chlorophyll if you want to purify it. You can take one teaspoon in a glass of water five to eight times a day if you feel it is necessary.

Hepatitis

Greens, carrying a great amount of iron, are one of the aids for hepatitis. Each organ has its own affinity for certain elements in order to be well-balanced. The liver, as we said, is an iron organ and needs lots of it because detoxification takes place here.

Acute cases of hepatitis, even where liver damage has occurred, have shown improvement with liquid chlorophyll.

In cases of hepatitis, chlorophyll does not have to come in concentrated form. All greens, and especially the Vegetable Top Liquor, will be of benefit. Dandelion tea which is rich in iron and potassium, is another fine chlorophyll source.

If you use the concentrated liquid, use one teaspoon in a glass of water three times a day.

Moving the Bile Along

When a chronic underactive bile has developed, liquid chlorophyll may also help get the bile flowing properly.

Bile is made up of the impurities taken out of the blood by the liver. Bile acts as an irritant to the bowel to get it moving.

For The Heart

Chlorophyll has been used as a remedy for heart trouble. One tablespoon of liquid chlorophyll in a glass of water three times a day with a teaspoon of honey is beneficial. This is used in conjunction with vitamin E, as much as 800 to 1200 units daily.

The heart and the arterial system that feed the various parts of the body need a lot of iron. When we have the proper amount we can keep a high blood count which is always a relief to the heart.

Regulating Menstruation

One of the greatest uses of liquid chlorophyll is for regulating the menstruation.

A woman came to me who had been menstruating almost every day for three months. An operation was advised. We put her on a heavy liquid chlorophyll diet, using two-three teaspoons in a glass of water five or six times a day. Within four days, the bleeding and hemorrhaging ceased. She has had regular menstrual periods since. Through this treatment and diet she regained her health and never had to have that operation.

My wife, Marie, helps me in the office and she can hardly believe how excited some of our woman patients get when they come in and say, "You don't know what a wonderful thing you have done for me!"

The seeming "miraculous" effect can be explained. Recognize again the fundamental truth that greens control calcium in the body. The menstrual blood has 40 times as much calcium as the regular blood. When you lose this calcium, there are many times liquid chlorophyll can help replenish it.

Chlorophyll Has Vitamin K

I am sure women are not meant to lose this calcium through menstruation. Further, here in chlorophyll is a rich source of Vitamin K — the anti-hemorrhagic vitamin. This vitamin was originally discovered in alfalfa juice.

Even for hemophilia, the condition where bleeding cannot be stopped, liquid chlorophyll may bring on a beneficial result, lessening the loss of blood which might have occurred without its additon to the diet.

Diabetes/Asthma

Diabetes and asthmatics are my most difficult cases. I almost walk out the door when I see them coming because of the reactions these patients have. Many times you can't control the reactions with juices or food. The patients get to a place where only drugs can handle them and, of course, I don't handle my patients with drugs.

If a patient is having an asthmatic attack, you can't say, "Well, take this glass of carrot juice. It will do you a lot of good!"

The carrot juice is a little late.

In cases of diabetes, improving such conditions demands a constant watch of the blood sugar and urine. The addition of chlorophyll will in the long run allow the patient to cut down on insulin, Orinase, and other drugs.

I went back East once and gave a talk on chlorophyll, showed some pictures of my garden, and discussed my other activities. There were about 35 men there, including doctors, attorneys, a priest, and other professionals. The priest said it was the finest sermon he ever heard.

One of the doctors came up and put his arm on my shoulder and said, "You know, ten years ago I would have called you a quack, but, I need you. My whole family is sick."

He wanted to learn more about chlorophyll.

Recently I received a letter from him. He described a case of one doctor at his clinic who had been using insulin for 16 years but was off of it after using liquid chlorophyll and proper nutrition for only three months.

Still further, a doctor in San Diego came to me with a blood sugar of 380 which is an extreme case. A normal blood sugar level is between 80 and 120.

In three months time, his blood sugar level was down to a high normal of 118. Does he think a miracle has occurred! He doesn't travel without his liquid chlorophyll. He reports he is maintaining his blood sugar level with proper nutrition and chlorophyll.

Balancing Milk Drinking

If you don't eat the balance of foods to keep up the chemical balance the body needs, and if by the end of each day the iron content has been low, in the long run you'll become anemic. The people who drink a lot of milk and don't eat a lot of the iron-producing foods are likely candidates for anemia. They are the ones with the poor bloodstream which is unable to get rid of all the catarrh that is produced by the excess amount of milk drinking.

Wherever we find milk drinking people who also have a lot of the natural iron-rich foods, we never find they are anemic, and we never find they have any catarrhal problems.

Milk Can't Build A Good Bloodstream

You can't build a good bloodstream on milk. You can get much calcium from it for rebuilding and rejuvenating and for knitting bone structure, but a child can grow all right and still produce many other diseases caused by anemia.

Milk and milk products become anemic-producing because we use so much of them (about 25% of our diet are milk products; it should be 6%) and because it has such little iron in comparison to the body's needs. One way to make up for the poor iron content of milk is to add greens to the diet. Greens are high in iron.

Green salads, tops of vegetables, and supplemental forms of liquid chlorophyll will help balance milk. One teaspoon of liquid chlorophyll in a juice or in the milk itself will also help.

Babies Need Greens

Raw milk is almost a complete food, but only for babies. Babies are given enough iron in their liver to last through the first year of life. After this period, they need greens and vegetables also. If a child is raised on milk, cheese, ice cream, and buttermilk, take a green food and see that it is eaten along with the milk product.

One day someone is going to wise up and produce a green cheese and maybe a yogurt with liquid chlorophyll in it.

Improving Milk Production

Chlorophyll can be used to increase milk production in a nursing mother. In Japan, when a mother is going low on milk, doctors recommend more greens.

When a nursing mother tells me she doesn't have enough milk, I know she is not eating enough salads or drinking green juices. A nursing mother can have all the milk she wants to produce by eating greens in most cases.

Greens keep up the milk supply in goats and cows . . . why not human mothers?

Sores

Any sore responds well to liquid chlorophyll, whether the sore is inside the body or outside. No sore heals without calcium, which is called the knitter in the body. Because liquid chlorophyll controls calcium, it is highly recommended when you want any knitting to take place. A good nutritional program which will furnish the other elements is also recommended.

Chlorophyll as Deodorizer

Liquid chlorophyll is a great deodorizer. When goats have the proper amount of greens they don't have the odors around them.

In 1950, Dr. Westcott experimented to prove that when chlorophyll is taken internally it reduces and could eliminate body odor. It could be effective for as long as 18 hours with doses of 65 to 200 mgs.

One of the many phenomena of Nature is that at the time fawns are born to deer, the fields are covered with lush green vegetation of which the mother partakes before giving birth. As a result, these fawn are born with a natural protection, which makes them odor-free for at least four days. This makes them safe from other predators which might be attracted to the smell of the new-born deer until the fawn can run and fare for itself.

Bacteria Fighter

Of special concern to healing wounds is that the chlorophyll furnished by green vegetables is a "wonder food." Bacteria cannot live in it. Theodore M. Rudolph in his book *Chlorophyll, Nature's "Green Magic,"* points out that "Reduction of pus formation and tissue destruction can be speedily accomplished with the application of a chlorophyll ointment as soon after the accident as possible."

Pains from inflamation and swelling caused by bacteria infection can be reduced by such application, and healing will soon follow.

Pyorrhea

In pyorrhea, the gums and tooth structure are attacked by old decaying material. Use papaya tablets in the upper and lower parts of the jaw on each side of the mouth. Let them dissolve slowly for 15-20 minutes and then wash your mouth with a liquid chlorophyll.

As A Nasal Douche

Use liquid chlorophyll as a nasal douche to clean out the nasal passages. Dilute it first with water and then sniff into the nose. This is very helpful with many catarrhal drainage problems, inflamed nasal passages, or a nasal drip. Use the same proportion of liquid chlorophyll as you do for a gargle, ½ teaspoon to ½ cup of water.

This country has gone crazy deodorizing people. These odors exist because we do not eat right. What we eliminate through evaporation from the skin are odors that come from the 61 flavors of ice cream and the 75 varieties of pickles. These are not natural foods and end up in a toxic state in the body!

Using liquid chlorophyll is an aid for cutting down the need of underarm deodorizers.

When you have a clean body, these odors do not exist.

Bad Breath

Bad breath usually comes when the stomach is not operating properly. Taking a drink of liquid chlorophyll in the morning is a good way to start the day.

½ teaspoon in ½ cup of water of the concentrated liquid chlorophyll is the suggested amount.

Another suggestion for getting rid of bad breath is as follows: After you have eaten garlic or onions you can free your breath of the odors by chewing vigorously on a small amount of fresh, uncooked parsley. Be sure to swallow the masticated parsley after chewing. Bad breath has formed the basis of thousands of lines of advertising matter. Parsley is a safe, satisfactory remedy to be used in an emergency. A liquid chlorophyll capsule will also do a good job.

As a Gargle

Chlorophyll makes an excellent gargle. My wife gets a slight irritation in her throat every once in a while and she finds liquid chlorophyll a tremendous help. We have had cases where people had inflamed tonsils and found relief. It is also excellent when you have had tooth surgery.

For sore throat liquid chlorophyll is recommended three times a day using ½ teaspoon to ½ cup of water.

For inflamed tonsils use the same amount.

For tooth surgery use one teaspoon in ½ cup of water or use straight concentrated liquid chlorophyll. Work the liquid around the teeth and hold in the mouth for two-three minutes. It can then be swallowed as the gargle can.

Ulcers of The Stomach

Ulcers of the stomach have a great friend in liquid chlorophyll, especially when added to flaxseed tea. The ulcer patient should take three cups of flaxseed tea a day and sometimes even five cups a day with the addition of a teaspoon of liquid chlorophyll to each cup.

The stomach can receive green packs by combining greens with flaxseed tea which will coat the stomach with its thickness and lubricating ability. Use a teaspoon of concentrated liquid chlorophyll to a cup of flaxseed tea.

In cases of peptic ulcers, gastric ulcers, and gastritis, use one tablespoon of liquid chlorophyll in a half glass of warm water every hour or two for three-four days. Then cut the amount and time down as needed.

For a Bleeding Bowel

A mixture of flaxseed tea with liquid chlorophyll is a good remedy for a bleeding or ulcerated bowel; however, in cases of any hemorrhaging consult your doctor.

Making Flaxseed Tea

There are two methods I recommend for making flaxseed tea. Take two tablespoons of flaxseed and place in a pint of water. Boil the seeds for ten minutes, then turn off the heat, and let the liquid stand. Strain off the seeds when it cools down. Discard the seeds and use the liquid only.

The second method is to add one quarter cup of flaxseed to one quart of warm water and let stand overnight, straining off the seeds before drinking. Do not boil. This second method preserves more of the Vitamin F, and because it is more natural, is more soothing and healing to the bowel membranes.

Use one teaspoon of liquid chlorophyll to a cup of flaxseed tea and take three times a day.

Chlorophyll Enemas

Chlorophyll enemas using flaxseed tea are also wonderful for clearing irritations of the lower bowel, colitis, ulcerations, painful hemorrhoids, and piles.

I have also used chlorophyll successfully in my diet work in enemas for cleansing the colon and as a deodorant of the bowel wall. Whatever osmosis takes place in the bowel will certainly pick up the materials found in chlorophyll because these chemicals found in this pure green water are the purest fluids we have.

I use these enemas in extreme cases I must deal with, and do not feel the average person in good health must have them to clean the bowel.

Hemorrhoids

For rectal problems, especially hemorrhoids, place a chlorophyll capsule in the rectum, a vitamin A capsule, and a garlic oil capsule.

Left in overnight, this combination provides a good healing agent for the rectal tissues.

Colitis

Combine one teaspoon Aloe Vera gel to a cup of flaxseed tea and add ½ teaspoon of liquid chlorophyll concentrate. Take this by mouth and by enema. The enema should be retained for 10-15 minutes before expelling. This remedy is soothing and heals inflamed tissues of the bowel, stomach, and gastro-intestinal tract. (See "Bleeding Bowel" for recipe for flaxseed tea.)

Many doctors recommend whey because of its high sodium content for conditions of peptic and duodenal ulcers. Add liquid chlorophyll to the whey and it will help healing even further.

Discharges

Liquid chlorophyll is indicated for any catarrhal, phlegm, or mucous discharge from any part of the body—ear, nose, bronchial tubes, etc. When toxemias throughout the body are cleaned up, discharges go away.

When liquid chlorophyll is used internally, it becomes a dietetic treatment and can clear up such conditions. Healing takes place from the inside outward and in most cases an internal rather than an external treatment is best. With chlorophyll, however, surprising results occur both ways.

Varicose Veins

Varicose Veins and leg troubles need extra chlorophyll because of stagnant blood circulation and poor exchange of venous blood in the lower extremities. A good circulation forces stagnant blood out of the lower limbs but we ALSO need the good blood which liquid chlorophyll can provide to revitalize the vascular system in the legs.

As a Pain Killer

Chlorophyll will relieve pain caused by inflammations. However, you can't take it and expect it to act like an aspirin. As I have already said, when you make food your medicine, you find that cells will regenerate and repair. Nature does the curing, but she is slow.

CHAPTER EIGHTEEN

A CHLOROPHYLL CREDO: "WHEN YOU ARE GREEN INSIDE, YOU ARE CLEAN INSIDE"

For many years I have stressed: "When you are green inside, you are clean inside." To be free from disease and disharmony in this present and coming survival age, we must be clean.

Food Should Flow Through Us

"Clean Inside" means eating foods which flow through us, leaving their amino-acid proteins, enzymes, minerals, vitamins, and other vital energies. This food should not stop to putrefy and ferment.

Waste materials not regularly evacuated from the intestines will accumulate. Their toxic products of decomposition may be taken up by the blood and lymph and be carried to every cell in the body. When decomposing food stagnates, the body must handle these poisons at great expense of energy. This energy could be put to better use for body repair, rejuvenation, and creative accomplishment.

Plenty of Friendly Bacteria

"Clean inside" means having a healthy ratio of acidophilus to bacillus colli bacteria in the bowel. This healthy ratio helps maintain

121

a good acid-alkaline balance in the digestive channels, a balance which avoids acid-caused toxic conditions throughout the body. When the body is highly acid, a host of conditions including many chronic diseases can exist.

Cleaned-Out Diverticula

"Clean inside" means having diverticula (pockets) that do not have a lot of toxic material in them. These pockets hold gas-producing, low-grade infections which feed the bloodstream poisons the body reabsorbs.

Toxic wastes from diverticula promote ill-health in all inherently weak organs where they throw off the functioning ability.

A Healthy Bloodstream

"Clean inside" means a well-working bloodstream. This organ must carry food from the digestive tract to the cells, oxygen from the lungs to the cells, and waste products away from cells to the organs of excretion.

The bloodstream regulates body temperature, distributes the hormones made by the ductless glands, helps maintain the proper balance between acid and alkaline throughout the body, and builds immunity to unfavorable germ life. All this must be accomplished by a clean bloodstream free of abnormal levels of toxic poisons.

The bloodstream must be well-supplied with health promoting minerals-such as iron, oxygen, copper, cobalt, etc.

Development of abnormal tissue in organs can be the result of this toxic-laden bloodstream that has built these conditions.

A Well-Nourished Liver/Thyroid

"Clean inside" means having a liver capable of detoxifying the impurities which run through it. It must be nourished by vibrant red blood cells, well supplied with iron and oxygen which give the ability to burn waste materials.

We need a thyroid gland free from toxins and able to provide oxygen to the body. This gland must be supplied with iron and iodine. It must be free of toxins to function properly.

Raw Salads

"Green inside" results from an active regimen of raw salad eating, salads with plenty of green leafy chlorophyll-rich vegetables.

Raw salads are high in the vital minerals iron, copper, cobalt, potassium, magnesium, etc. - elements crucial to build the blood, muscles, and bowel tissue.

Raw salads are high in digestive enzymes. These make it easier to digest the heavier foods we eat, such as starches and proteins. Enzymes are necessary to keep the lifestream flowing through us. Enzymes help all body functions.

Raw salads have a nice fibre which keeps the food in the bowels moving along.

Raw Chlorophyll Juices

"Grean inside" results from an active regimen of drinking raw chlorophyll-rich juices. These juices maintain the 80-90% liquid environment which our body demands so our cells can be fed properly and life processes maintained.

These juices, as do salads, provide food for the acidophilus bacteria culture in our bowels.

Chlorophyll juices provide the finest distilled water our body can have, a water which cleans our tissues of drug deposits and artificial sprays.

Chlorophyll juices are charged with chemical elements which build the red blood cell count. Because of the heavy iron in chlorophyll, it makes life easier by letting us take oxygen from the air. Chlorophyll feeds our cells with many trace minerals difficult to get any other way.

Chlorophyll juices, rich in iron and oxygen, help detoxify the liver.

Chlorophyll juices, because they bring oxygen to the thyroid gland and keep this gland working at the healthy rate of speed, are a blessing to this, our vital body servant.

Because the charged chemical elements come to us in chlorophyll, a substance chemically kin to our own blood composition, they can be absorbed into our body with little loss of vital energy. This saving of energy is crucial for any healing to take place.

Chlorophyll Supplements

"Green inside" means using chlorophyll supplements when we must. Chlorophyll liquid concentrates and chlorophyll juice in tablet form are available in the health market place.

Herb Seasonings

"Green inside" results from using life-giving herb seasonings instead of liver embalming, high-heat produced salt. Pepper, 20 times more irritating to the liver than alcoholic drinks, can also be replaced.

Herb Teas

"Green inside" results from using health-giving herb teas and not liver-hardening heart stimulating coffee. Coffee creates acids in the body, develops a toxic liver in time, and kills off our friendly acidophilus bacteria, one of its most destructive effects.

Herb/Vegetable Broths

"Green inside" results from using herb and dried vegetable broths, and dehydrated vegetable powders. These broths are available commercially. Herbs can be grown in the house or harvested in the field and then blended into broths.

SURVIVAL LAW NO. 12

The more greens - vegetables, herbs, grasses, weeds, juices, broths, etc., we can get into our body on a regular basis, the cleaner, purer, stronger, and healthier our body will be.

Sprouting Program

"Green inside" comes when we master a program of sprouting seeds and grains. These sprouts provide soft bulks for the intestines which aid digestion. They also have valuable enzymes.

These sprouts also provide natural minerals and vitamins in plentiful proportions to feed every organ in the body.

Sprouts are food so alive that they are still "screaming" when we chew them; it is the aliveness of raw foods which keeps us alive!

Sprouts are pure and clean. This makes them one of our crucial survival foods in this day when unfertilized and unsprayed vegetables are hard to get.

The Wild Weeds

"Green inside" is ours when we know our edible wild grasses and weeds; when we let our imagination flow in the kitchen. What herb dressings, seasonings, and herb vinegars can we create to liven up our meal tonight?

Put Nature to Work for You

Greens, holding chlorophyll and other life-preserving vitality, are one of Nature's greatest healers because they are Nature's greatest body cleanser and builder; herein lies a Natural Law from Nature's manual, a law of infinite meaning.

Chlorophyll is not medicine as practiced by a medical doctor, but it is medicine as practiced by a nutritionist. I use food for the correction of ailments and I DO correct ailments through nutrition. I prove it every day. I can't say that we cure, but many conditions have been improved for the better using different types of foods. No one should object to pure, clean, whole foods.

I recognize that anything new is automatically opposed. When you start telling people what you believe in, and it may sound just a little bit different than the average cereal flakes that they are having for breakfast or the white bread for lunches, then you are considered a little bit of an oddball. You are considered a little bit off.

Well, possibly I have some advanced ideas for us to consider. Those people I have cared for prove I am on the right path. This isn't a kid's job. I have spent 45 years in this work, and I have thousands of case histories behind me.

I should have learned something if I had never gone to school. I should have learned from just experience, and I tell you my observations and experience with people tell me that liquid chlorophyll is one of the finest substances you can add to your diet.

I was inspired to use this green fluid from the Bible story of Daniel. I was inspired to use the tops of vegetables—the Vegetable Top Liquor—to chop up the leaves very finely. When I put these cut leaves into water they began to bleed into the water. I have strained off this water and fed it to patients and seen wonderous things happen. They came back fairer than all the King's youth.

I believe in Nature, and in her greens and grasses of the field. I believe that God gave them to us. I believe the further we can get from a drug store and the closer we can get to God and Nature, the better our chances for a long, healthy, and beautiful life.

Recognize, however, that chlorophyll is only one side of the health story. Don't go out of this book and turn green. I just don't want you to become a jolly GREEN giant. I want you to become a JOLLY giant. You have to make a balanced health package for yourself.

There are some people who go overboard from a juice standpoint and take far too much. Yet, on the other hand, the average person doesn't know how it feels to feel wonderful. People go around half-dead and think they're alive. Wow, what a world we would live in if everybody had a perfect blood count and all the chemical elements they needed for good health.

I am an observer. Many times I say, "I believe," or many times I say, "It is my opinion." One of these days what I tell you here will become all FACT, but I have to wait for science to come along.

The day is coming when we will realize health magic through chlorophyll.

APPENDIX A: FRIENDS FROM THE GARDEN & FIELD

The list of *Friends from the garden and field* is intended as a beginners guide to some of the greens which should play a part in our diet. Begin to familiarize yourself with the following greens and master the use of each. Advanced lessons on the planting and care of many of them will be included in future lessons of this *Magic Survival Kit*.

ALOE VERA—This wonderful member of the lily family can grow in many places. It has a muscilaginous liquid helpful for skin burns and rashes and offers quick relief for muscle cramps when applied locally. Use internally for inflamed bowel, colitis, and ulcers of the stomach. Aloe Vera also has outstanding healing qualities for —hemorrhoids and piles when used in enemas. (See the section on cactus for remedy details.)

ALFALFA—Alfalfa in the form of sprouts is one of the purest foods we can have for the body. They promote a healthy bowel activity. Alfalfa leaves in the form of alfalfa tablets are helpful in keeping the pockets and other areas in the stomach free of stagnant purtrefactive materials. The alfalfa teas are excellent sources of alkalinity.

BEET GREENS—Beet greens, containing potassium, magnesium, iodine and iron, are an excellent body mineralizer. Clean and wash thoroughly. Use stems if they are tender. Cut up fine and steam like spinach. Beet green juice is wonderful for the liver and gall bladder. A half glass per day will also help bowel movements.

BOK CHOY—This Chinese cabbage is high in sulphur and iron, and because of its rich green content, is also high in potassium. Sulphur purifies and activates the body and tones the system.

BRUSSELS SPROUTS—This member of the cabbage family is also high in sulphur. Many people find it gas-forming in a raw juice; however, cooked cabbage will cause less gas.

CABBAGE—Cabbage juice is excellent for the stomach. It contains sodium, the youth element, which is stored there and also potassium, the muscle toner. A medical doctor at Stanford cleared up ulcers of the stomach using cabbage juice as a remedy.

Use raw cabbage in salads. There is more vitamin C in a half cup of raw cabbage than in a medium sized orange.

CELERY GREENS—Celery greens should be juiced along with the stalk. The potassium in the green tops balances the high sodium content of the stalk. If a person just takes the stalk he will be getting a concentrated form of sodium. The overabundance of sodium causes water to be held in the body. An overabundance of potassium can also cause problems, but few people ever get too much potassium because our body requires so much of it. It specifically feeds the muscle structure which makes up 80% of our body.

CHIVES—These are high in potassium, calcium and sulphur. Serve in salads and with cottage cheese. Chives are good for catarrhal elimination.

COMFREY—Comfrey juice is a fine remedy for ulcers of the stomach, is an aid for catarrhal conditions of the lung structure, and is beneficial for leg ulcers. Drink comfrey and fenugreek tea two or three cups a day for catarrhal conditions in the lungs. For leg ulcers, make an external pack. Chop fresh comfrey leaves finely (or use comfrey tea leaves). Mix them with water to hold them together and then put the mixture on the body, placing a damp cloth around it. This damp cloth should be covered with a dry cloth. Leave it on all night.

DANDELION GREENS—Although we find that all the greens are excellent for the liver, dandelion greens act especially well on it and the gall bladder. These greens are fine sources of calcium, manganese, chlorine, potassium, and iron. Use in salads.

ENDIVE—While endive is bitter, it is also excellent for reducing and makes a nice bulk for the bowel. Wash and serve in salads.

ESCAROLE—Here is another lettuce family member which is high in potassium. Add fruit juice to make it more palatable.

GREEN KALE—Here is one of the highest greens in calcium. It also belongs to the sulphur family.

MALVA—This wild weed grows heavily throughout the United States and elsewhere. One pound has 50,000 units of natural vitamin A which is so important in clearing up infections. Malva is tasty in salads and can also be steamed like spinach. Be sure to clean the leaves first, using a teaspoon of purex to a gallon of water. Soak the leaves for five minutes and then wash them off.

NASTURTIUM FLOWERS—These make another source of greens. Have them in your salads.

PARSLEY—Parsley makes an outstanding chlorophyll drink for the kidneys. You can dry it and make a parsley tea. Always use in soup and broth. A pinch of it every day will help to keep the kidneys in good order.

PEPPERMINT, SPEARMINT, MINT—These herbs are high in chlorophyll. They make excellent flavorings for bitter drinks. They are wonderful when it comes to driving gas out of the intestinal tract and they also make the tract smell sweet.

SPINACH—Try to have a little bit in a raw juice but do not overdo it because it contains oxalic acid which interferes with the absorbtion of calcium in the body. Swiss chard and beet greens also contain this acid—so does chocolate. Considering the wide variety of greens in the garden, spinach should be eaten about once a week.

SWISS CHARD—A brother of the spinach, this family member is a little more palatable than its kin. It also contains oxalic acid.

TURNIP GREENS—I have referred to Dr. Goldstein's contribution to health of the South when he recommended using these greens to control calcium metabolism in the body. When Dr. Goldstein used this remedy most symptoms of pellagra disappeared.

WATERCRESS—Here is one of the greatest of our garden friends for trimming off weight. Anyone overweight should consider getting more potassium and less sodium in the diet. Here is a high potassium food. Sodium holds water while potassium helps to get rid of it. Because it grows along streams, it is seldom sprayed.

WHEATGRASS JUICE—This is one of the finest health builders I know. (See details under section on it.)

This list is by no means complete but is intended as a reminder of some of the valuable foods we have in the garden. I hope you begin using them if you are not already doing so. (For additional vegetable ideas see Dr. Jensen's *Vital Foods for Total Health.*)

HOW TO CONSUME VEGETABLE AND FRUIT
JUICE—Vegetable and fruit juice drinks can be of great benefit
only if they are consumed properly. They must be drunk slowly,
almost eaten, to aid the saliva in mixing with the juice. The
blender or other juicing technique grinds up the vegetable in a
way the human teeth would grind them, but it does not furnish
the saliva to mix with the juice! The thorough salivation of the
liquefied food against the roof of the mouth to get rid of all the
air unavoidably whipped into it is essential, or it will cause much
discomfort from gas in the stomach and intestines. Juices must
never be consumed like drinking water, even though the
broken down particles are largely water. There are still some solid
food particles contained in the juice, and these must be mixed
with the natural saliva before they are swallowed. Digestion
begins in the mouth.

Another important factor to consider in liquefication of food
is that oxidation occurs very rapidly. We should begin drinking the
juice as soon as it is made. Liquefied food spoils quickly unless
consumed within a reasonably short period of time. It should
not be kept for more than six hours. If liquefied food is to be
stored, it must be filled to the top of the container so that no
air remains to unite with the juice.

COOKING VEGETABLES—Unquestionably raw foods were
absolutely meant for man to eat. There is no denying that they
are the best and contain everything as man should have it, but
we know that we do not live as natural people any longer.
Many elements are destroyed in cooking. These may be replaced
by drinking raw vegetable juices and eating a raw salad with each
meal. Cook vegetables as little as possible to save all the vitamins
and chemical elements, and keep a very low fire under them.
Consider steaming your vegetables in a pot with tight lid and

very little water so that they may steam in their own juices.

VEGETABLE SOUPS—Liquefied vegetables and fruits need not be served only as cocktails or salads. They also make delicious soups—raw, warmed or cooked. A parsley, green pepper, and summer squash combination with some added herbs, warmed a little and seasoned with butter, can be served as a soup.

SOUPS FROM SPROUTS—Every household should have a sprouting program—grains, seeds, legumes, for their high nutritive value. Then these should be processed in the liquefier with a suitable liquid or water and a little carrot juice. Add some sliced leek, a few green vegetable leaves, vegetable broth powder, oil, and some slippery elm powder (to give it a smooth consistency and to benefit the stomach). When a warm soup is desired, heat only the liquid used as the base.

HOMEMADE BROTH AND SOUP—There is nothing quite like a homemade broth or soup, and with the blender to aid you, there are simply thousands to create. Do not use meat stock except in special broths for specific conditions. Use vegetable stock, herb teas or bran water. A broth may be made from any vegetable or combination of vegetables with vegetable powder seasoning instead of common salt.

Barley and Green Kale Soup

(Best Calcium and Winter Soup)
½ cup Barley—soak in water overnight.
1 qt. water
2 stalks celery, chopped finely
1 onion, chopped finely
Add. Bring to boil and simmer 40 minutes.
2-3 cups Kale, chopped finely
2 tsps. Vegetable Seasoning
Add and cool further 20 minutes.
Parsley, finely chopped
A little butter
Some Raw cream
Add as about to serve.

Vital Broth
(excellent for elimination)
from Dr. Jensen's *Vital Foods for Total Health.*
   2 C. *carrot tops*
   2 C. *celery stalk*
   2 C. *beet tops*
   ½ *tsp. Savita*
   2 C. *potato peeling (cut ½-inch thick)*
   2 C. *celery tops*
   2 *qts. Sparkletts distilled water*

Add a carrot or onion for flavor if desired.

Finely chop or grate vegetables. Bring slowly to a boil, simmer approximately twenty minutes. Just use broth after straining.

(For additional soup ideas, see Dr. Jensen's *Vital Foods for Total Health, Blending Magic,* and *Soup's On).*

CHLOROPHYLL SOUPS—Liquid chlorophyll in concentrate can be added to any soup when you are about to serve it. It will not throw the taste of the soup off. At least a teaspoon to a cup of broth or a bowl of soup is the suggested amount. You can add raw greens to any of the bean and squash soups. Celery juice makes a fine addition at the time the soup is being served.

A SPECIAL BROTH—My special broth for those patients who need the greatest help getting out of their sickness is made of liquefied beets, carrots, string beans, and soy milk powder. Liquefy the vegetables in some water, strain the juice out, and then add the soy milk powder to the juice. Discard the pulp.

BUYING VEGETABLES—Buy a variety of vegetables every day. Find out when the fresh vegetables are delivered to your market and get them as fresh as possible. Many tasty tops of vegetables are thrown away, such as beet and turnip tops, and these are worth asking for. When buying lettuce or leafy vegetables, choose the greenest as they are highest in chlorophyll. The green outside cabbage leaves have 40% more calcium than the bleached out leaves inside which do not reach the sun. Use the tops of vegetables the first day. The underground vegetables will last longer. Feed greens to your chickens and get orange egg yolks instead of pale yellow.

The salad is no better than the greens that go into it. Salad greens should be garden-fresh and free from blemishes and sprays. They should be carefully washed in cold running water, then shaken gently to remove the excess moisture. Slight dampness helps to keep the greens fresh and crisp while in the storage tray, but too much moisture shortens the storage life. Place the vegetables in either the refrigerator-hydrator in a clear plastic bag, or wrap them in a clean white cloth until ready to use. When oil is added to the greens before refrigerating, the leaves are coated and this discourages wilting. Lemon juice, or the remainder of the dressing, may be added when the salad is ready to serve.

Select only fresh young greens. *Celery* should have clean, thick stalks with fresh, green leaves. Avoid limp stalks and brown tops. *Cabbage* should be heavy for its size, with the head firm and crisp. *Chives* should be bright green, and in tufts. *Chicory* leaves should be broad, with curly edges. *Beet tops* should be fresh and small, from young beets. *Endive* should be curly, in bunchy green heads with yellow-green centers. Belgian type has smooth, yellow-white, elongated heads. *Escarole* is like endive, but has broader leaves and may be either blanched or green. *Lettuce* comes in large, soft heads in different varieties, some of which are Boston, butter, leaf and Romaine. Most types have tender, delicate leaves, but Romaine has long, wide leaves of dark green, and has a more sturdy texture. Both the inner light green tender leaves, as well as the outer dark green leaves, are good for salad. The darker, tougher, stronger flavored greens such as escarole, chicory, parsley, watercress, romaine, etc., are richer in minerals and vitamins and this makes up for their lack of delicate flavor. From *Salad Favorites Around the World,* by Dr. Jensen.

RAW SALADS—Raw salads cannot be taken in some intestinal tracts. Good vegetables and raw vegetable juice taken together, however, has almost the same nutritional benefits as eating the raw salads alone, and in many cases will not cause as much gas.

FREEZING VEGETABLES—One of the best ways to store vegetables is to freeze them. You can store all vegetables this way and can save yourself money by putting away vegetables and eating them out of season. Pea pods sell for $.50 a pound when they are fresh, but up to $3.00 a pound during the winter. Dr. Jensen's *Home Freezing* has much to say about this subject. subject.

JUICING VEGETABLES—If you have an intestinal tract which rebels against some of the nice vegetables, consider juicing them and avoiding the pulp. Most any vegetable can be juiced. Carrot greens, curly cabbage, swiss chard, and green kale sometimes fall into this category as do many vegetables grown on some soils or vegetables left in the ground a long time. Most of the chlorophyll is still available to you if you juice these rough vegetables.

ANOTHER METHOD FOR EXTRACTING CHLOROPHYLL—Another way to get the chlorophyll from greens is to put them in a blender with water. Turn to high speed until well-blended. Put the blended mixture in a muslin bag and squeeze as much juice out as possible. A good grinder can be substituted for the blender. Another solution is using a potato ricer. Simply put the pulp in the ricer and press as you would for ricing potatoes.

GREEN ICE CREAM—There is no reason in the world for not having green chlorophyll ice cream. Flavor homemade ice creams made from health recipes with mint, and ice cream will have a wonderful taste and odor. You can color it green with chlorophyll and the children will think it is a mint green ice cream.

> *Fresh Mint Freeze*
> (from Dr. Jensen's *Blending Magic*)
> ½ C. honey
> 1 C. water
> 1 C. whipped cream
> 1 C. fresh mint leaves

½ C. lemon juice
A little chlorophyll for coloring

Heat honey and water until honey is well dissolved. Add mint leaves to this and blend in blender on low speed until finely ground. Cover and let stand until cool. Stir in lemon juice and add green coloring. Strain into ice tray, discarding mint pulp and freeze until mushy. Turn into chilled bowl and beat with electric or hand beater until smooth. Fold in whipped cream and freeze until firm. Makes about three servings.

White Velvet Whipped Cream
½ C. warm water
4 Tbsps. soy milk powder
Blend on Medium.
½ C. safflower oil

—add gradually, continuing to blend slowly. Add a few drops pure vanilla and honey to sweeten. Blend until smooth "whipped" cream is obtained.

CHILDREN'S DRINKS—To create green drinks that even children will enjoy, mix about one-third green drink with apple, berry, or papaya concentrates and water or with soy milk powder. Pineapple, cherry, or black rasberry concentrates or mint will also sweeten the drink. Anytime you make a health milk shake, add the greens to balance out the small iron content which is found in the milk products.

SALAD TIPS—Salads present the best opportunity to get chlorophyll to your bloodstream, yet various greens in salads are difficult to take. Steamed raisins, sweet salad dressings such as cream cheese and blue cheese, nut butters and avocado dressing are ways to get around this problem. The finest salad is one with a rainbow of colors. There is much chlorophyll in other vegetables besides the greens. Have at least one salad a day, and two if possible.

IRON-RICH TONIC DRINK—When you are working with green vegetable juices which are on the bitter side and therefore distasteful, think of adding black grape or black cherry concentrate to the undiluted green juice. By adding an eggyolk, you create one of the finest high-iron protein drinks you can put into your body.

PRUNE JUICE—For a nerve and blood-building food, add prune juice to your green vegetable juice. Prunes are a good source of nerve salts.

SOY MILK—An inexpensive yet outstanding protein, soy milk powder can be added to bitter green vegetable drinks to quell the harsh taste. Adding soy milk to a vegetable broth will make the broth more easily digestible.

NUT AND SEED BUTTERS—When trying to gain or maintain health, give the body the easiest nutrients to digest. Whole nuts take from four to five hours to get through the stomach, and many times are not chewed well enough to get the nutritional value from them. Grind seeds and nuts and use in butters. Both sesame seed milk and almond nut milk drinks made from these butters are fine non-catarrhal producing drinks. Nuts and seeds are one of the best all-around foods for obtaining the trace minerals from the earth. These drinks and nut and seed butters will soften the taste of a broth which is heavy in greens. The greens also will make an excellent addition to such drinks.

SALT—Table salt should be dropped from the diet. We do not need it when we have plenty of greens everyday. To change from it, use vegetized salt purchased in your health food store. Vegetable concentrates in powdered form and herbs are also excellent seasonings.

Dr. Dicky in South Africa found that he could trade for anything he wanted from the Pygmies with salt. However, when he began giving them plenty of green vegetables, their desire for salt lessened and he could no longer bargain with salt alone because it had lost its value to them. Still further, we find that deer are not attracted to salt licks when they have the greens of the field. It is during the dry grass season, when the chlorophyll is lacking, that they desire this salt. Liquid chlorophyll contains the most potent, vital cell salts a person can take into the body. When we crave salt, I feel it is because the body doesn't have all the elements it should have from these greens.

AS A PRESERVATIVE—Concentrated liquid chlorophyll is a natural preservative. I recommend virgin cold-pressed olive oil because it has chlorophyll in it. It doesn't go rancid or spoil as the second or third pressings of the olive oil. You can improve your diet and avoid oil rancidity by adding a little extra chlorophyll to the salad oil you are using.

HERB TEAS—If an herb tea mixture is composed mostly of roots and bark, the tea can be boiled in order to get the most of the value out of it. If it is composed mostly of leaves and flowers it should not be boiled as boiling will evaporate the aromatic properties of the tea. Herbs can be used for teas in their green stage, but it is a common practice to dry them first.

Eight heaping teaspoonsful of dry herbs may be used when one ounce liquid tea is stipulated, or one teacupful of dry herbs may be used to make one pint of tea.

DECOCTION: This is made by boiling herbal or vegetable material a certain length of time, usually from ten to twenty minutes. The boiling should be done slowly in a closed vessel. Where exact proportions are not stated on the label, use one teaspoonful of the herb to make one cup of tea and drink two or three cups a day when needed.

INFUSION: The herb is added to boiling water. The vessel is then taken off the fire, covered, and the contents allowed to steep until lukewarm, then strained. As ordinary tea is made. Use ½ to 1 teaspoonful of whole or chopped herb to make one cup of tea. Use ¼ to ½ teaspoonful of powdered herb. Never boil an infusion. Steep only, 15 to 30 minutes.

If an herb tea is to be taken for a long time, it is best to take it alternately for three days, omit for three days and continue again for three days and so on until the desired results are obtained.

HERB VINEGARS—Vinegar has long been a favorite medium in which to preserve the aromas of sweet and savory herbs. Tarragon, basil, burnet, etc., lose much of their aromatic oil content in the drying process, but the aroma is well preserved in vinegar. In times far gone by, vinegars were also aromatized with rose petals, elder, rosemary, and other spicy, scented blooms. These flower vinegars redeem the most uninteresting salad and sauce, and until we have used them we cannot realize how plain store vinegar is. In making, use good apple cider vinegar. Take the container with you to the herb patch and pack the jar full of succulent tips and leaves as fast as you pick. Press them down hard in the vinegar, seal, and allow to stand on the sunny shelf a week or more. Then strain and add to the liquid more fresh herbs if you like, but the jar can also be left as it is and the vinegar-soaked leaves chopped in salad dressing the winter through. The whole process is begun on a warm summer morning when the herbs are "juicy" with vigorous growth.

A mixed herb vinegar is useful. The herbs which might be used are lemon balm, marjoram, thyme, basil, tarragon, chive, savory, burnet, and a bit of rue. Let no one strong herb overpower another. Herbs most frequently used alone are basil, tarragon, burnet, and mint.

MAKING MINT VINEGAR—Bring to a boil one quart apple cider vinegar. Add one cup raw sugar, a pint of spearmint leaves, and young stem tips. Stir and crush. Boil for a few minutes. Strain and bottle hot in glass jars. This is one of the best flavors for iced tea and fruit punches when fresh mint leaves are not at hand. It is also used as the base for mint sauce.

Tarragon vinegar is excellent with chicken salad, one teaspoon to one cup of mayonnaise. Garlic vinegar and olive oil on sliced beefsteak tomatoes need no other garnish.

HERB PUNCHES—Use grape juice, orange, or lemonade as a base. For sweetening, honey is ideal with herbs, or make a syrup of boiled raw sugar to which an unpeeled lemon or orange has been added at the beginning. To this add lemon thyme, as much as you like, mints, or lemon balm. Either strain, or if the syrup is bottled for future use, leave in the herbs. Spearmint is the best julep herb and is used without crushing. The stimulating flavor of borage tips and flowers has always been recognized in claret drinks. Always use clean herbs free from water in these beverages. The tips of young shoots are the most pungent.

HERB BUTTERS—Follow these steps to make herb butters:

(1) To one-half cup of soft raw unsalted butter add one teaspoon minced clove of garlic, fresh rosemary, and a dash of fresh lemon juice. Mix well and let stand at room temperature for two hours. Use with hot green beans.

(2) To one-half cup of soft butter add one teaspoon finely minced parsley and one-eight teaspoon crushed dried tarragon.

(3) To soft butter add one-eighth teaspoon garlic salt, one-half teaspoon each of minced fresh or dried thyme, marjoram and rosemary. Spread on bread and serve with soup or salad. Delicious.

(4) To one-half cup soft butter add one-half teaspoon each minced marjoram, thyme and parsley. Mix lightly through hot, cooked rice and let stand, keeping hot for fifteen minutes before serving. Try buttering hot corn-on-the-cob with this butter.

HERB SAUCE—In two tablespoons olive oil, gently cook half a cup each chopped onion and green pepper until soft. Sprinkle with half teaspoon dried sweet basil and pour in one can tomato juice (or sauce). Use with Spanish omelet. Stir and cook on low heat for about fifteen minutes.

HERB CHEESES—Mix into one pint small-curd cottage cheese the following: one tablespoon olive oil, one teaspoon each finely minced parsley, marjoram, thyme, and a shake of garlic salt. Refrigerate overnight in glass or china bowl. Serve on rye bread or as a salad with fresh tomatoes. Grate one-half pound very sharp, aged cheddar cheese into a bowl. Add one-half cup grape juice and one teaspoon crumbled dried sage leaves. Mix well, let stand at room temperature for an hour or two to blend.

APPENDIX C: JUICES FOR SPECIFIC AILMENTS

Vegetable juices can be assimilated quickly and easily. In fact, the green, or chlorophyll, is taken up by the blood with little effort. We have seen a million red blood cells added to a blood count in a month and a half by adding vegetable juices to the diet.

There is no better organic water for your system than raw vegetable juice. For best results, have at least a pint a day with meals or between meals.

Many times diarrhea is experienced while taking vegetable juices. In most cases it is necessary, as Nature is doing a little housecleaning. To bathe the tissues in your body with these juices will, in time, cleanse them and then rebuild, rejuvenate, and feed a starved body.

Make these cocktails in proportion pleasing to the taste. Usually any juice mixed with an equal proportion of another juice will be about right. We find carrot, celery, and parsley good in nearly all conditions and good in combination. They mix with any fruit or vegetable juice and can be had between meals, with most any meal, and any time during the day. However, when you mix fruit juices with vegetable juices, as pineapple and tomato, it is best not to mix with starches. Acid fruits and starches do not combine. Vegetable juices go best with starches, and all starches should have vegetable juices with them. When having carrot juice, drink with a straw and mix well with saliva.

Disorders in the Body	*Health Cocktail Suggestions*
Anemia	Parsley and Grape Juice
Asthma	Celery and Papaya Juice
Bed Wetting	Celery and Parsley Juice
Bladder Ailments	Celery and Pomegranate Juice
Catarrh, Colds,	
Sore Throat	Watercress and Apple Juice. (Add ¼ teaspoon of Pure cream of Tartar)

Constipation, Stomach	Celery and a little sweet cream.
Ulcers	Spinach and Grapefruit Juice.
Colds, Sinus Troubles	Celery and Grapefruit Juice. (Add ¼ teaspoon of pure Cream of Tartar.)
Colitis, Gastritis, Gas	Cocoanut Milk and Carrot Juice
Diarrhea, Infection	Carrot and Blackberry Juice
Fever, Gout, Arthritis	Celery and Parsley Juice
Gall Bladder Disorders	Radish, Prune, Black Cherry, and Celery Juice
General Housecleaning	Celery, Parsley, Spinach, and Carrot Juice
Glands, Goitre, Impotence	Celery Juice. 1 teaspoon Wheat Germ, and 1 teaspoon Nova Scotia Dulce
Heart Disturbances	Carrot and Pineapple Juice and Honey
High Blood Pressure	Carrot and Parsley Juice and Celery
Indigestion, underweight	Cocoanut Milk, Fig Juice, Parsley and Carrot Juice
Insomia, Sleeplessness	Lettuce and Celery Juice
Kidney Disorders	Celery, Parsley and Asparagus Juice
Liver Disorders	Radish and Pineapple Juice
Nervous Disorders	Radish and Prune Juice and Rice Polishings
Nerve Tension	Celery, Carrot and Prune Juice
Neuralgia	Cucumber, Endive and Pineapple Juice
Nerve Quieter	Lettuce and Tomato Juice
Overweight, Obesity	Beet Green, Parsley and Celery Juice
Poor Circulation	Beet and Blackberry Juice
Poor Complexion	Cucumber, Endive and Pineapple Juice
Poor Memory	Celery, Carrot and Prune Juice and Rice Polishings
Poor Teeth	Beet Greens, Parsley and Celery Juice and Green Kale
Reducing	Parsley, Grape Juice and Pineapple Juice
Rheumatism, Neuritis Neuralgia	Cucumber, Endive, and Goat's Whey

Rickets   Dandelion and Orange Juice
Scurvy, Eczema   Carrot, Celery and Lemon Juice

The juices suggested for a specific body disorder may be used separately or in combination with each other.

(From Dr. Jensen's *Vital Foods for Total Health*)

DR. JENSEN'S BALANCED DAILY EATING REGIMEN

Make a habit of applying the following General Diet Regimen to your everyday living. This is a healthy way to live *because*, when followed, you do not have to think of vitamins, mineral elements or calories. I will give you more specific instruction for your troubles after you have made this daily regimen automatic.

The best diet, over a period of a day, is two different fruits, at least four to six vegetables, one protein and one starch, with fruit or vegetable juices between meals. Eat at least two green leafy vegetables a day. 50% to 60% of the food you eat daily should be raw. Consider this regimen a dietetic law.

Rules of Eating

1. Do not fry foods or use heated oils.
2. If not entirely comfortable in mind and body from the previous meal time, you should miss the next meal.
3. Do not eat unless you have a keen desire for the plainest food.
4. Do not eat beyond your needs.
5. Be sure to thoroughly masticate your food.
6. Miss meals if in pain, emotionally upset, not hungry, chilled, overheated, and during acute illness.

Impositions For Getting Well

Learn to accept whatever decision is made.

Let the other person make a mistake and learn.

Learn to forget and forgive.

Be thankful and bless people.

Live in harmony — even if it is good for you.

Do not talk about your sickness.

Gossip will kill you. Don't let anyone gossip to you either. Gossip that comes through the grape vine is usually sour.

Be by yourself every day for ten minutes with the thought of how to make yourself a better person. Replace negative thoughts with uplifting, positive thoughts.

Skin brush daily. Use a slant board daily.

Have citrus fruits in sections only, never in juice form.

Have only a limited amount of bread (with a lot of bowel trouble, no bread).

Exercise daily. Keep your spine limber. Develop abdominal muscles. Do sniff breathing. Have a daily set of exercises.

Grass walk and sand walk for happy feet.

No smoking, drinking, spitting or cussing. Keep away from "spitty" people.

Bed at sundown, 9 p.m. at the latest, if you are at all tired, fatigued and unable to do your work with vim and vigor. If you are sick you must rest more. Sleep out of doors, out of the city, in circulating air. Work out problems in the morning, don't take them to bed with you.

Food Healing Laws

1. *Natural food*—50% to 60% of the food eaten should be raw.
2. *Your Diet should be 80% alkaline and 20% acid.* Look at the acid—alkaline chart in *Vital Foods for Total Health*, page 100.
3. *Proportion*—6 vegetables daily, 2 fruits daily, 1 starch daily and 1 protein daily.
4. *Variety*—vary sugars, proteins, starches, vegetables and fruits from meal to meal and from day to day.
5. *Overeating*—you can kill yourself with the amount of food you eat.
6. *Combinations*—separate starches and proteins. One at lunch and the other at supper. Have fruits for breakfast and at 3:00 p.m.
7. *Cook without water*—Cook without high heat. Cook without air touching hot food.
8. *Bake, broil or roast*—If you eat meat, have it. Have lean meat, no fat, no pork. Use unsprayed vegetables if possible and eat them as soon after picked as possible.
9. *Use stainless steel, low-heat cooking utensils*—It is the modern health-engineered way of preparing your foods.

Before Breakfast

Upon arising, and one-half hour before breakfast, take any natural, unsweetened fruit juice, such as grape, pineapple, prune, fig, apple or black cherry. Liquid chlorophyll can be used—take 1 teaspoonful in a glass of water.

You can have a broth and lecithin drink if you desire it. Take 1 teaspoonful of vegetable broth powder and 1 tablespoonful of lecithin granules and dissolve in a glass of warm water.

On doctor's advice you may have citrus fruits such as orange, grapefruit, lemon or tomato.

Between fruit juice and breakfast, follow this program: Skin brushing, exercise, hiking, deep breathing or playing. Shower. Start warm; then cool off until your breath quickens. Never shower immediately upon arising.

Breakfast

Stewed Fruit, One Starch and *Health Drink* or . . . *Two Fruits, One Protein* and *Health Drink*. (Starches and health drinks are listed with the lunch suggestions.) Soaked fruits, such as unsulphured apricots, prunes, figs. Fruit of any kind—melon, grapes, peaches, pears, berries or baked apple, which may be sprinkled with some ground nuts or nut butter. When possible, use fruit in season.

Suggested Breakfast Menus

MONDAY

Reconstituted Dried Apricots, Steel-Cut Oatmeal—Supplements Oat Straw Tea. Add Eggs, if desired. or Sliced Peaches, Cottage Cheese—Supplements, Herb Tea.

TUESDAY

Fresh Figs, Cornmeal Cereal—Supplements, Shave Grass Tea. Add Eggs or nut butter, if desired. or Raw Apple Sauce and Blackberries, Coddled Egg — Supplements, Herb Tea.

WEDNESDAY

Reconstituted Dried Peaches, Millet Cereal—Supplements Alfamint Tea. Add Eggs, cheese or nut butter, if desired or Sliced Nectarines and Apple Yogurt—Supplements, Herb Tea.

THURSDAY

Prunes or any reconstituted dried fruit, Whole Wheat Cereal—Supplements Oat Straw Tea or Grapefruit and Kumquats, Poached Egg—Supplements, Herb Tea.

FRIDAY

Slices of Fresh Pineapple with Shredded Coconut, Buckwheat Cereal—Supplements Peppermint Tea or Baked Apple, Persimmons, Chopped Raw Almonds, Acidophilus Milk—Supplements, Herb Tea.

SATURDAY

Museli with Bananas and Dates, Cream—Supplements Dandelion Coffee or Herb Tea.

SUNDAY

Cooked Applesauce with Raisins, Rye Grits—Supplements Shave Grass Tea or Cantaloupe and Strawberries, Cottage Cheese—Supplements, Herb Tea.

Preparation Hints

Reconstituted dried fruit—Cover with cold water, bring to boil and leave to stand overnight. Raisins may just have boiling water poured over them. This kills any insects and eggs.

Whole grain cereal—To cook properly with as little heat as possible, use a double boiler or thermos-cook your cereal.

Supplements—(Add to cereal or fruit) Sunflower seed meal, rice polishings, wheat germ, flaxseed meal (about a teaspoonful of each.) Even a little dulce may be sprinkled over with some broth powder.

10:30 A.M.

Vegetable broth, vegetable juice or fruit juice.

Lunch

Raw Salad, or as directed, *One* or *Two Starches*, as listed and a *Health Drink*.

Note: If following a strict regimen use only one of the first seven starches daily. Vary the starch from day to day.

Raw Salad Vegetables: Tomatoes (Citrus). Lettuce (Green leafy type only, such as romaine). Celery, cucumber, bean sprouts, green peppers, avocado, parsley, watercress, endive, onion(s), cabbage(s) are sulphur foods.

Starches

1. Yellow corn meal. 2. Baked potato. 3. Baked banana (or at least dead ripe). 4. Barley—a winter food. 5. Steamed brown rice or wild rice. 6. Millet—have as a cereal. 7. Banana squash or Hubbard squash.

Steel cut oatmeal, whole wheat cereal, Dr. Jackson's meal, whole grain, Roman meal, shredded wheat bread (whole wheat, rye, soy bean, corn bread, bran muffins, Rye Krisp preferred.)

Drinks

Vegetable broth, soup, coffee substitute, buttermilk, raw milk, oat straw tea, alfamint tea, huckleberry tea, papaya tea, or any health drink.

Suggested Lunch Menus

MONDAY
Vegetable Salad, Baby Lima Beans, Baked Potato, Spearmint Tea
TUESDAY
Vegetable Salad - with health mayonnaise if desired. Steamed Asparagus, Very ripe bananas, or Steamed, Unpolished Rice, Vegetable Broth or Herb Tea.
WEDNESDAY
Raw Salad Plate, Sour Cream Dressing, Cooked Green Beans, Corn Bread and/or Baked Hubbard Squash, Sassafras Tea
THURSDAY
Salad - French Dressing, Baked Zucchini and Okra, Corn-on-cob, Rye Krisp, Buttermilk or Herb Tea.
FRIDAY
Salad, Baked Green Pepper, stuffed with eggplant and tomatoes, Baked Potato and/or Bran Muffin, Carrot Soup or Herb Tea.
SATURDAY
Salad, Turnips and Turnip Greens, Baked Yams, Catnip Tea.
SUNDAY
Salad, Lemon and Olive Oil Dressing, Steamed Whole Barley, Cream of Celery Soup, Steamed Chard, Herb Tea.

Salad Vegetables: Use plenty of greens. Choose four or five vegetables from the following: Leaf lettuce, watercress, spinach, beet leaves, parsley, alfalfa sprouts, cabbage, young chard, herbs, any green leaves, cucumbers, bean sprouts, onions, green peppers, pimentoes, carrots, turnips, zucchini, asparagus, celery, okra, radishes, etc.

3:00 P.M.
Health cocktail, juice or fruit.

Dinner

Raw Salad, Two Cooked Vegetables, One Protein and a *Broth* or *Health Drink* if desired.
Cooked Vegetables: Peas, artichokes, carrots, beets, turnips, spinach, beet tops, string beans, swiss chard, eggplant, zucchini,

summer squash, broccoli (s), cauliflower (s), cabbage (s), sprouts (s), onion (s), or any vegetable other than potatoes.
Drinks: Vegetable broth, soup, or health beverage.

Proteins

Once a Week: Fish—use white fish, such as sole, halibut, trout or sea trout.
Vegetarians—Use soy beans, lima beans, cottage cheese, sunflower seeds and other seeds; also, seed butters, nut butters, nut milk drinks, eggs.
Three Times a Week: Meat—use only lean meat. Never pork, fats or cured meats.
Vegetarians—Use meat substitutes or vegetarian proteins.
Twice a Week: Cottage Cheese or any cheese that breaks.
Once a Week: Egg Omelet.
If you have a protein at this meal, health dessert is allowed, but not recommended. Never eat protein and starch together. (notice how they are separated.)
You may exchange your noon meal for the evening meal, but follow the same regimen. It takes exercise to handle raw food, and we generally get more after our noon meal. That is why a raw salad is advised at noon. If one eats sandwiches have vegetables at the same time.

Suggested Dinner Menus

MONDAY
Salad, Diced Celery and Carrots, Steam Spinach, waterless-cooked, Puffy Omelet, Vegetable Broth.

TUESDAY
Salad, Cooked Beet Tops, Steak, Broiled, or Ground Beef Patties - Tomato Sauce, Cauliflower, Comfrey Tea.

WEDNESDAY
Cottage Cheese, Cheese Sticks, Apples, Peaches, Grapes, Nuts, Apple Concentrate Cocktail.

THURSDAY
Salad, Steamed Chard, Baked Eggplant, Grilled Liver and Onions, Persimmon Whip (Optional), Alfa-mint Tea.

FRIDAY
Salad, Yogurt and Lemon Dressing, Steamed Mixed Greens, Beets, Steamed Fish - with slices of Lemon, Leek soup.

SATURDAY
Salad, Cooked String Beans - Baked Summer Squash, Carrot and Cheese Loaf, Cream of Lentil Soup or Lemongrass Tea, Fresh Peach Jello, Almond-Nut Cream.

SUNDAY
Salad, Diced Carrots and Peas, Steamed Tomato Aspic, Roast Leg of Lamb, Mint Sauce.

Vegetarians—Use vegetarian dishes in place of meat dishes.

APPENDIX E: A VITAMIN SURVEY OF GREENS

Green Vitamin Survey

Vitamin A—Conditions caused by lack of Vitamin A: Loss of weight and vigor, loss of vitality and growth, loss of strength, acne, poor vision, rough skin, diarrhea. Conditions controlled by Vitamin A: Makes tissues more resistant to colds and catarrhal infections in respiratory organs, promotes feeling of well-being. Sources: Green leafy vegetables and yellow vegetables. Spinach, swiss chard, green lettuce, cabbage, carrots, green peas, endive, beet leaves, mustard greens, brussel sprouts, celery, yellow squash.

Vitamin B—Conditions caused by lack of Vitamin B: Nervous exhaustion, loss of growth, loss of reproduction function, loss of appetite, intestinal gas, faulty nutrition and assimilation, soreness and pain, lack of digestive juices. Conditions controlled by Vitamin B: Makes for better absorption of food and normalizes the brain and nervous sytem by increasing metabolic processes. Sources: Asparagus, spinach, peas, turnip greens, mustard greens, chard, celery, carrots, cabbage, beet leaves, lettuce, broccoli, peppers, avocados.

Vitamin C—Conditions caused by lack of Vitamin C: Tender painful swelling of joints, poor health, faulty nutrition, scurvy, loss of appetite, loss of weight, irritable temper, poor complexion, loss of energy, irregular heart action, rapid respiration, reduced hemoglobin, reduces secretions of adrenals, cataract, hemorrhage. Conditions controlled by Vitamin C: a marvelous health promoter as it wards off acidosis. Sources: cabbage, spinach, peas, broccoli, rutabagas, collards, brussel sprouts, celery, parsley, endive, watercress, turnips, cucumbers, cauliflower, radishes.

Vitamin D—conditions caused by lack of Vitamin D: rickets, soft bones, lack of body tones, fatigue, respiratory infections, irritability, restlessness, constipation, prolapsus, dental caries, retards growth, instability of nervous system. Conditions controlled by Vitamin D: Facilitates absorption of calcium and phosphorus from foods, consequently a green bone builder. Guards against tuberculosis. Regulates mineral metabolism. Sources: green leafy vegetables grown in the sunshine.

Vitamin E—Conditions caused by lack of Vitamin E: Sterility, loss of adult vitality. Conditions controlled by Vitamin E: Essential in reproduction, poor lactation, menstrual disorders, miscarriage, dull mentality, pessimism, despondency and loss of courage. **Sources: Green, leafy vegetables and sprouts.**

Vitamin F—Conditions caused by lack of Vitamin F: stunted growth, sexual immaturity, falling hair, baldness, loss of appetite, skin disorders, nervousness and eczema. Conditions controlled by Vitamin F: Vitamin F is necessary for all-around development. Sources: Root vegetables and fresh spinach.

Vitamin G (B$_2$)—Conditions caused by lack of vitamin G: nervous disorders, irritability, pellagra, skin eruptions, loss of hair, stomach disorders, cataracts, old age, lack of growth, poor appetite, digestive disturbances. Conditions controlled by Vitamin G (B$_2$): Prolongs life span, increases adult vitality. Sources: Green leafy vegetables.

The foregoing survey is a sampling of the vitamin content of various greens. Other vitamin contents of specific greens are included where these greens are discussed.

INDEX

If You've Enjoyed Reading This Book . . .

Vibrant Health From Your Kitchen—One of Dr. Jensen's latest and greatest books. In this book, he teaches the basics of health and nutrition. A food guide for family health and well-being. The reader learns how proper foods can overcome certain mineral deficiencies, allergies and build immunity.

Food Healing of Man—Innumerable experiences are recounted in Dr. Jensen's work with both human beings and animals. The book is a comprehensive layman's guide to the healing power of foods elaborating on nutritional deficiencies. Lists 28 factors necessary for correcting body ailments. A study is made as to why foods heal and the reason for supplements.

Nature Has A Remedy—Solutions applying nature's restorative powers are discussed. A nature encyclopedia covering hundreds of ailments. Teaches methods of taking care of various symptoms encountered with diet, water treatments, physical exercise, climate, environment and others.

Tissue Cleansing Through Bowel Management—Toxic-laden tissues can become a breeding ground for disease. Elimination organs, especially the bowel, must be properly taken care of. This book tells the reader how. Bowel management through a balanced nutritional program with adequate fiber in the diet and regular exercise can often do wonders. A special 7-day cleanse will bring back energy, regenerate tissues and allow good food to let nature do its healing work.

The Healing Mind Of Man—The spiritual, mental and physical qualities of man must be considered for healing. All body functions depend upon our mind, and must be brought into balance before healing can occur.

A New Lifestyle For Health and Happiness—A concise summary of Dr. Jensen's most effective methods for restoring and maintaining good health. A sound program outlined with practical applications and daily charts for improving a person's lifestyle.

Vital Foods For Total Health, A Cook Book And Kitchen Guide—You are what you eat! Your health, your looks, and even the length of your life is affected by your diet. Your meals may look good and taste good, yet lack the vital elements your body requires. So to keep your health and your looks, or to regain them, eat correctly! This book will tell you how. This is a complete cookbook which combines health teaching and the newer knowledge of nutrition.

Foods That Heal—In the first half of this book, Dr. Jensen focuses on the philosophy and ideas of Hippocrates, the brilliant work of Dr. V. G. Rocine, and concludes with a look at his own pioneering work in the field of nutrition. The second half is a nutritional guide to fruits and vegetables.

For information on *Dr. Jensen's Food Products for the 21st Century* and for a *free* catalog of all his books and supplies, please write to:

World Keys to Health and Long Life

Based on Dr. Jensen's travels to over fifty-five countries, this fascinating book describes the health and longevity secrets of centenarians interviewed in the Hunza Valley of India; Vilcabamba, Peru; the Caucasus Mountains of the Soviet Union; and other places around the world.

Doctor-Patient Handbook

Discover the reversal process and healing crisis that Nature uses to rid the body of disease and restore well-being. Here is a fresh approach to wholistic health.

Slender Me Naturally

Dr. Jensen's answer to fad diets that don't work is a natural weight loss program that does. Developed over fifty-eight years of experience with overweight patients, this program is a healthful and effective way of losing unwanted weight.

Breathe Again Naturally

Get rid of asthma, allergies, bronchitis, hay fever, and other respiratory problems. Dr. Jensen discusses nutrition, herbs that work, food supplements, breathing exercises, attitude, and climate.

Arthritis, Rheumatism and Osteoporosis

Are you among the one in four Americans who suffers from arthritis, rheumatism, or osteoporosis? Would you like to know what to do about it? This book is for you.

Foods That Heal

This book presents the basic principles of Hippocrates, Dr. Rocine, and Dr. Jensen regarding the use of foods to help the body regain health. The author has also included a complete guide to the various fruits and vegetables we all need.

Beyond Basic Health

Dr. Jensen looks at the deteriorating state of modern man's health and offers practical advice and insights to those health professionals who must deal with today's devastating illnesses.

Love, Sex and Nutrition

Based on years of detailed study, this book explores the link between diet, sensuality, and relationships. This is an important and practical guide for people who wish to improve their sexuality safely and naturally.

For information regarding prices, write to:

Dr. Bernard Jensen
24360 Old Wagon Road
Escondido, CA 92027

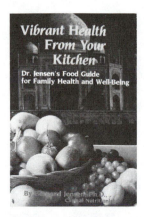